The Mass of a Photon

Josef von Stackelberg

The Mass of a Photon

Side Notes on Electrical Engineering

Springer

Josef von Stackelberg
Neusiß, Germany

ISBN 978-3-658-49407-0 ISBN 978-3-658-49408-7 (eBook)
https://doi.org/10.1007/978-3-658-49408-7

Translation from the German language edition: "Die Masse eines Photons" by Josef von Stackelberg, © Der/die Herausgeber bzw. der/die Autor(en), exklusiv lizenziert an Springer Fachmedien Wiesbaden GmbH, ein Teil von Springer Nature 2025. Published by Springer Fachmedien Wiesbaden. All Rights Reserved.

This book is a translation of the original German edition "Die Masse eines Photons," 2nd edition, by Josef von Stackelberg, published by Springer Fachmedien Wiesbaden GmbH in 2025. The translation was done with the help of an artificial intelligence machine translation tool. A subsequent human revision was done primarily in terms of content, so that the book will read stylistically differently from a conventional translation. Springer Nature works continuously to further the development of tools for the production of books and on the related technologies to support the authors.

This Springer imprint is published by the registered company Springer Fachmedien Wiesbaden GmbH, part of Springer Nature.
The registered company address is: Abraham-Lincoln-Str. 46, 65189 Wiesbaden, Germany

Foreword

Over the course of many years, various seemingly unrelated investigations emerged from the broader field of electrical engineering. The motivations and reasons for these studies always stemmed from practical questions that I needed to examine in depth as part of my professional activities, or when I sought to understand a relationship that was generally taken for granted.

For all statements that I do not claim as my own, I have attempted to provide sources, even if I initially formulated a statement from memory as I believed I had learned it during my vocational training and studies, or as I recalled reading it in a book during my professional career. While writing this book, I was not able to find sources for every statement. Therefore, I would like to identify the following derivations as my own findings:

- In Chap. 3 the derivation of gravitational force from high-frequency electromagnetic oscillations,
- In Chap. 4 the derivation of individual oscillations, photon mass, and the four fundamental forces of physics from the toroidal helix,
- In Chap. 5 the alternative explanation of the relativistic mass increase phenomenon,
- In Chap. 6 the resolution of permittivity and permeability and their replacement by the speed of light in vacuum,
- In Chap. 7 the derivation of the Fourier series for the truncated sine,
- In Chap. 8 the derivation of the alternative diode equation, in which the temperature in the exponent is moved to the numerator.

For all other considerations, I make no claim to original discovery. In particular, regarding the proposed solution for determining the coefficients of the interpolation polynomial in Chap. 9 and the introduction of the fundamental equation for straight graphs in the double logarithmic coordinate system in Chap. 10, I cannot imagine that these findings have not already been published previously.

In an age dominated by the omnipresent Internet and the information and misinformation it disseminates, I consider it entirely normal to cite an encyclopedia such as

Wikipedia in my work just as I would any other source, since in my view the error rate of Wikipedia is at least not significantly higher than that of other publications. Especially when I examine the contents of the present investigative reports, in which I am able to question some generally accepted truths, I occasionally found myself wondering, over the course of this work, which publications can truly be trusted.

In the course of these explanations, I have described the calculation steps in considerable detail because I want to give every reader the opportunity to follow my calculations—either to identify any errors that may have occurred despite multiple reviews and recalculations (even though I do not expect this), or to understand how the results are obtained and perhaps even to build upon them further. Readers who are not interested in the mathematical derivations may skip these sections and continue reading the subsequent text.

There is one teacher I would like to gratefully acknowledge in connection with the content of this book: Professor Dr. Dr. Wolfgang Halang, emeritus professor of Information Technology at the FernUniversität in Hagen, who accompanied me for many years during my studies and occasionally in private life, and who, at various points, decisively encouraged me to fully articulate an uncomfortable thought before dismissing it as unproductive. From this encouragement to pursue a path that initially seemed unfamiliar or incorrect to its conclusion, several of the following chapters emerged.

Josef von Stackelberg

Foreword, 2nd Edition

In 2023, I had the opportunity to present some of the book's findings as part of an e-seminar series organized by VDE/VDI. While preparing the presentations, I noticed in several sections of the book that the lines of reasoning were not fully developed, which sometimes made the content difficult to understand. The second edition aims to address and close these gaps.

Furthermore, after some consideration, I have come to the conclusion that the photon mass I assumed in the first edition was too high. Therefore, Sect. 4.2 contains a substantive correction in this regard.

Contents

1 Introduction

It is said that photons have no mass. The reasoning behind this statement is that otherwise, they would not be able to travel at the speed of light. The foundation for this is provided by Albert Einstein with his theories of relativity.

On the other hand, photons can be deflected by masses. This deflection can be calculated. An initial estimate of the deflection was provided by Johannes Soldner; later, Albert Einstein corrected Soldner's calculations, which contributed, among other things, to the development of his General Theory of Relativity. Soldner's calculations are revisited in Chap. 2.

This deflection leads to the conclusion that photons respond to gravity; gravity is a defining property of mass. The second defining property of mass is inertia.

From this deflection effect, another assumption can be derived: that gravity and electromagnetic fields are somehow interconnected, since photons are generally understood to be electromagnetic waves. A technical explanation for the connection between gravity and electromagnetic fields can be found, which is developed in Chap. 3 and with which, in Chap. 4, a new fundamental element is introduced, from which everything is composed, starting from photons, electrons, and quarks to all matter in the universe. If one continues along this line of thought, the other two fundamental forces of physics—the strong and the weak force—can also be explained with this new fundamental element. The appealing aspect of this fundamental element is that it does not require leaving the three spatial dimensions and the time dimension.

Furthermore, in Chap. 5, the model is used to provide an explanation for why photons travel at the speed of light and under what conditions any given mass can also be accelerated to the speed of light or even beyond.

Finally, the further considerations in Chap. 6 even lead to the result that, through a simple re-evaluation of the concept of mass, two currently well-established constants in

J. von Stackelberg, *The Mass of a Photon*, https://doi.org/10.1007/978-3-658-49408-7_1

electrical engineering can be eliminated or replaced by the constant of the speed of light in vacuum.

All these lines of thought arise from critically examining existing laws of electrical engineering, as they do not deliver the desired results in daily practice, and from these critical examinations, new insights are derived.

This is why this book bears the subtitle, “Side Notes on Electrical Engineering,” because all the findings described here stem from practical investigations during my work as an electrical engineer.

Since, along the way, not only were insights gained that fit into the argumentation chain described above, but also other inconsistencies were identified and addressed, I am including these additional findings in the following chapters to make them accessible to the wider public as well.

The Fourier series of a sinusoidal voltage, for example, chopped by a thyristor, does not yield a chopped sinusoidal voltage, as is shown in Chap. 7.

The diode equation, described in Chap. 8, is a typical example of how, long ago, a mathematical model of a technical relationship was developed and subsequently no one dared to question this model, even though it is known that the existing model does not work.

2 Gravitational Lenses

Gravitational lenses are masses that bend light rays [1; 2]. This deflection occurs everywhere, but it is so slight that it becomes significant only in the combination of very large masses and great distances, since large masses produce correspondingly large transverse accelerations and great distances lead to measurable transverse displacements even for small deflection angles. These deflections of a light ray on its journey through the vastness of space mean that a star is not actually located where we appear to see it, but can be situated elsewhere entirely—a phenomenon popularly referred to as the "curvature of space."

A simple example is a light source located behind the Sun, which we can nevertheless observe because the rays from the source are deflected strongly enough by the Sun.

The first known calculation of this deflection was published by Johannes Soldner in March 1801, a mathematician from Franconia [3].

2.1 The Deflection of Light by Masses

In his work, Soldner arrived at a result that yielded exactly half the angle that was later verified by measurements.

Reconstructing the calculation described in the aforementioned source is not straightforward in all respects, as the terms for physical quantities, as we know them today, are in some cases assigned quite differently:

"Let the acceleration due to gravity at the surface of the body be g": However, Soldner does not specify the gravitational acceleration g in m/s^2, but rather as a numerical value to be multiplied by the radius of the Earth and by pi:

$$g_{\mathrm{EarthSoldner}} \cdot \mathbf{r}_{\mathrm{Earth}} \cdot \pi = g_{\mathrm{Earth}} \tag{2.1}$$

J. von Stackelberg, *The Mass of a Photon*, https://doi.org/10.1007/978-3-658-49408-7_2

Similarly, he expresses the speed of light as follows:

$$\mathrm{v}_{\mathrm{EarthSoldner}} \cdot \mathrm{r}_{\mathrm{Earth}} \cdot \pi = \mathrm{c}_{\mathrm{Vacuum}} \tag{2.2}$$

with

$$\mathrm{g}_{\mathrm{EarthSoldner}} = 5.75231 \cdot 10^{-7} \frac{1}{\mathrm{s}^2}$$

$$\mathrm{r}_{\mathrm{Earth}} = 6.369514 \cdot 10^6 \mathrm{m}$$

$$\mathrm{v}_{\mathrm{EarthSoldner}} = 15.562085 \frac{1}{\mathrm{s}}$$

If these values are substituted into his formula for the deflection of a light ray

$$\tan \omega = \frac{2 \cdot \mathrm{g}}{\mathrm{v} \cdot \sqrt{\mathrm{v}^2 - 4 \cdot \mathrm{g}}} \tag{2.3}$$

and ω is expressed in degrees instead of radians (by dividing the radian value by ($\pi \cdot 180$)), then according to his work, one obtains a deviation angle of 10^{-3} arcseconds.

For the sake of completeness, it should be mentioned that this deviation angle describes only the deviation of the light ray from the light source to the Earth. Naturally, there is an equally large deviation when the light ray passes the Earth and continues on to a more distant observer.

While Soldner provides the initial values for his calculation of the deflection of light rays by the Earth, no such initial values are given for the deflection of light by the Sun. He only states the result as $\omega = 0.84''$, which is approximately half the deflection determined by measurement.

To reproduce this result, the dimensions for the Sun [4]

$$r_{Sun} = 6.967 \cdot 10^8 m$$

$$g_{Sun} = 273.8 \frac{m}{s^2}$$

are used to determine the two values $\mathrm{g}_{\mathrm{Sun\ Soldner}}$ and $\mathrm{v}_{\mathrm{Sun\ Soldner}}$ according to Eq. 2.1:

$$g_{Sun\ Soldner} = 1.2509437 \cdot 10^{-7} \frac{1}{s^2}$$

$$v_{Sun\ Soldner} = 0.1370647 \frac{1}{s}$$

If these are again substituted into the formula for tan ω, the result is

$$\omega = 2.7469296''$$

If this value is divided by π, one obtains

$$\omega = 0.87437485'',$$

which is strikingly similar to the value published by Albert Einstein in his first attempt and is half the value he later corrected [5]. The correction is logical, since the original value describes only half the path of the light ray (see above).

In any case, all previous calculations have been carried out conventionally, without taking time dilation or length contraction into account.

The deflection of a particle moving at the speed of light can also be estimated as follows (Fig. 2.1). In this approach, the trajectory taken by the photon as it passes the deflecting mass is divided into a finite number of segments, each with the length of the radius r of the deflecting mass. At each intermediate point after a segment, the force and motion conditions are considered, and the deflection angle for the subsequent trajectory is calculated based on the values of the path already traveled. The deflection angle α is defined as the angle between the direction of motion at the source and the new direction of motion from the intermediate point under consideration. The analysis fundamentally relies on the direction of the attractive force exerted on the photon by the deflecting mass. This force results in an acceleration a pointing in the same direction, from which the transverse acceleration a_q, perpendicular to the direction of motion, can be derived using trigonometric functions. From the transverse acceleration, the transverse velocity v_q in the direction of the transverse acceleration and the transverse displacement s_q can be calculated.

$$\tan \alpha_0 = \frac{x_h}{d - x} = \frac{r}{d} \tag{2.4}$$

with

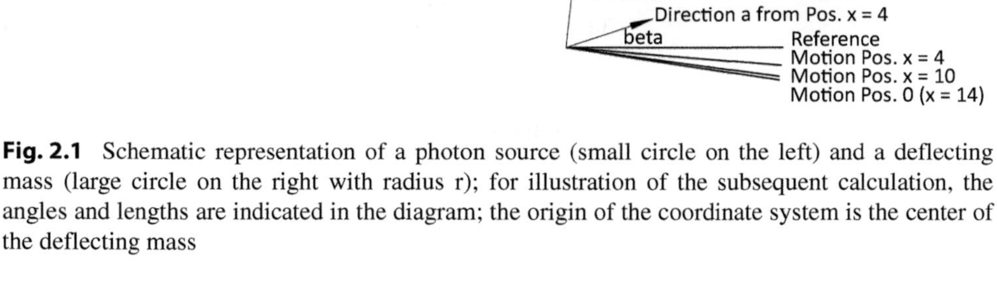

Fig. 2.1 Schematic representation of a photon source (small circle on the left) and a deflecting mass (large circle on the right with radius r); for illustration of the subsequent calculation, the angles and lengths are indicated in the diagram; the origin of the coordinate system is the center of the deflecting mass

$\alpha_0 =$ Initial angle between the line connecting the two centers and the tangent at position $x \to d$

$x_h =$ Distance between the line connecting the two centers and the tangent at position x; x_h is perpendicular to the line connecting the two centers and depends on x

$d =$ Distance between the two centers

$x =$ Variable for the distance between the center of mass and the position under consideration; $x \in [0; d]$

$r =$ Radius of the deflecting mass

From this, one can derive

$$x_h = (d - x) \cdot \frac{r}{d} \tag{2.5}$$

for the position $x = d$.

Moving along the connecting line from $x = d$ toward $x = 0$, x_h decreases by the amount of the photon's transverse displacement s_{q-}.

$$x_h = (d - x) \cdot \frac{r}{d} - s_{q-} \tag{2.6}$$

While

$$\alpha_0 = \arctan \frac{r}{d} \tag{2.7}$$

applies,

$$\alpha = \arcsin \frac{s_{q-}}{r} \tag{2.8}$$

Furthermore,

$$\tan \beta = \frac{x_h}{x} \tag{2.9}$$

applies, with

$\beta =$ the angle between the line connecting the two centers and the radial line from the center of mass to the point where x_h meets the tangent; β depends on the position x

$$\tan \beta = \frac{(d - x) \cdot \frac{r}{d}}{x} \tag{2.10}$$

$$\beta = \arctan \frac{d \cdot r - x \cdot r}{x \cdot d} \tag{2.11}$$

$$\gamma_0 = \alpha_0 + \beta \tag{2.12}$$

$$\gamma_0 = \arctan\frac{r}{d} + \arctan\frac{d \cdot r - x \cdot r}{x \cdot d} \tag{2.13}$$

using the addition theorem for arctan x + arctan y

$$\frac{r}{d} \cdot \frac{r \cdot (d - x)}{d \cdot x} = \frac{r^2}{d^2} \cdot \frac{d - x}{x} < 1 \quad for \quad x \in [r; d] \tag{2.14}$$

one obtains

$$\gamma_0 = \arctan\frac{\frac{r}{d} + \frac{d \cdot r - x \cdot r}{x \cdot d}}{1 - \frac{r}{d} \cdot \frac{d \cdot r - x \cdot r}{x \cdot d}} \tag{2.15}$$

Accordingly, the following applies over the entire range of x

$$\gamma = \alpha + \beta \tag{2.16}$$

and

$$r_{mm} = \frac{x}{\cos\beta} \tag{2.17}$$

For

$$F_g = \frac{g \cdot m_1 \cdot m_2}{r_{mm}^2} \tag{2.18}$$

with

$g =$ Gravitational constant $= 6.67408 \cdot 10^{-11}$ m3/(kg $\cdot$ s^2)
$m_1 =$ Mass of the attracting body in kg
$m_2 =$ Mass of the photon
$r_{mm} =$ Distance between the centers of mass of the two bodies in m

and

$$F = m_2 \cdot a \tag{2.19}$$

with

$m_2 =$ Mass of the photon
$a =$ Acceleration caused by the force

By equating the forces, one obtains

$$a = \frac{g \cdot m_1}{r_{mm}^2} \tag{2.20}$$

On the one hand, it is evident that the photon mass cancels out, i.e., the deflection is independent of the mass of the deflected object, which is consistent with Newtonian mechanics (see also Newton [6]); on the other hand, this acceleration yields the transverse acceleration a_q of the photon, which depends on the angle γ (the angle dependent on the position x) and on the angle α:

$$a_q = a \cdot \frac{\sin \gamma}{\cos \alpha} \tag{2.21}$$

$$a_q = \frac{g \cdot m_1}{r_{mm}^2} \cdot \frac{\sin \gamma}{\cos \alpha} \tag{2.22}$$

where

$$r_{mm} = \frac{x}{\cos \beta} \tag{2.23}$$

From this, the transverse velocity v_q and the transverse displacement s_q can be derived:

$$v_q = a_q \cdot t + v_{q-} \tag{2.24}$$

$$s_q = \frac{a_q}{2} \cdot t^2 + v_{q-} \cdot t + s_{q-}, \tag{2.25}$$

where v_{q-} and s_{q-} are to be taken from the previous time interval.

It can be assumed as an approximation that

$$t = \frac{r}{c}. \tag{2.26}$$

For the subsequent α, then

$$\alpha = \arcsin \frac{s_{q-}}{r} \tag{2.27}$$

If one fills an Excel spreadsheet with this recursive series, one obtains at the position of the deflecting mass a deflection angle of 0.87534285 arcseconds (Fig. 2.2).

This estimate was made under the assumption that photons move at the speed of light and possess mass, i.e., that they respond to the gravitational field of the deflecting mass.

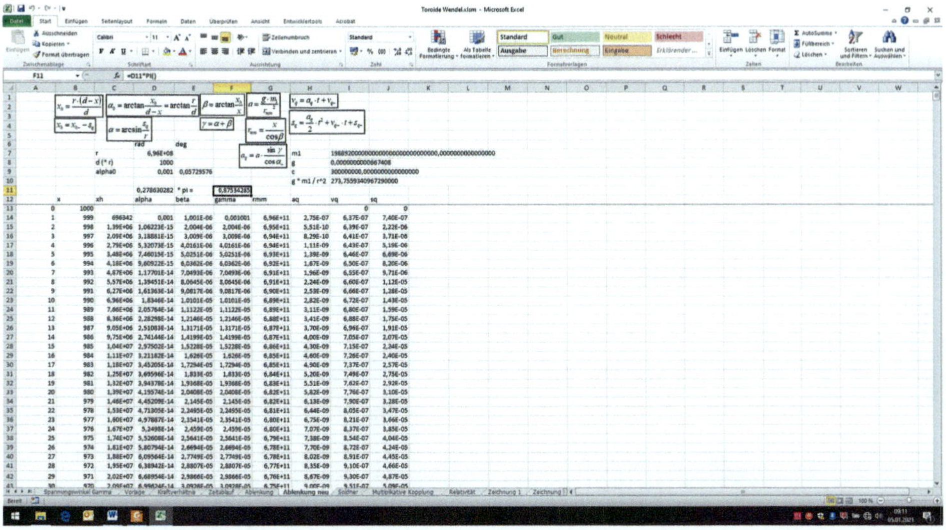

Fig. 2.2 Excerpt from the Excel spreadsheet showing the numerical determination of the deflection angle of a light beam by a mass

References

1. Wikipedia, Gravitationslinseneffekt, 1. 3. 2021
2. S. Hawking; Das Universum in der Nussschale; Kober, Deutscher Taschenbuch Verlag, 2004
3. wikisource.org: Joh. Soldner; Über die Ablenkung eines Lichtstrals von seiner geradlinigen Bewegung; Berlin; 1801
4. Wikipedia, Sonne, 1. 3. 2021
5. P. Schneider, J. Ehlers, E. E. Falco; Gravitational Lenses; Springer Verlag Berlin; 1999
6. Jay O'Rear; Physik; Carl Hanser Verlag München; 1989

3 The Gravitational Field

Unlike electric and magnetic fields, which surround bodies and can exert both attractive and repulsive forces on corresponding bodies, gravitational fields between two bodies, according to current understanding, always act attractively on each other. If one assumes that gravitation is an electromagnetic effect, then the two corresponding fields of two bodies would need to have the following properties in order to always act attractively: the product of the two fields would have to be unipolar on average over time. The following derivation explains how a force that is unipolar on average can arise from alternating fields.

The force between two bodies surrounded by electric fields is described as follows:

$$F_{el} = -\frac{q_1 \cdot q_2}{4 \cdot \pi \cdot \varepsilon_0 \cdot d^2} \tag{3.1}$$

If both charges q_1 and q_2 are either positive or negative, the force between these two charges is negative, i.e., repulsive. Different charges, therefore, result in a positive force, which is attractive.

A corresponding description of the force between two bodies with mass is given by

$$F_{grav} = \frac{g \cdot m_1 \cdot m_2}{d^2} \tag{3.2}$$

Mass values are always positive, and therefore the force between them is always positive, i.e., attractive.

There is no corresponding formula describing the magnetic force between two bodies with magnetic fields, because magnetic fields always have closed field lines, i.e., they have neither a source nor a sink.

Now, if we assume that two structures generate alternating fields over time that interact with each other, and that this interaction should always be attractive for both

J. von Stackelberg, *The Mass of a Photon*, https://doi.org/10.1007/978-3-658-49408-7_3

structures—that is, the mean value of the interaction is nonzero—this can be derived mathematically as follows:

$$W = F_1 \cdot F_2 \tag{3.3}$$

with

$$F_1 = a_1 \cdot \sin(\alpha_1 + \phi_1) \tag{3.4}$$

and

$$F_2 = a_2 \cdot \sin(\alpha_2 + \phi_2) \tag{3.5}$$

with

$F_n =$ field
$a_n =$ amplitude of the field
$\alpha_n =$ time-dependent angle of the field
$\varphi_n =$ time-independent phase angle of the field

This yields

$$W = a_1 \cdot \sin(\alpha_1 + \phi_1) \cdot a_2 \cdot \sin(\alpha_2 + \phi_2) \tag{3.6}$$

$$W = a_1 \cdot a_2 \cdot \sin(\alpha_1 + \phi_1) \cdot \sin(\alpha_2 + \phi_2) \tag{3.7}$$

It must be examined under which conditions the multiplication of two sinusoidal fields that mutually influence each other results in a net force over time. Such a net force over time occurs when, in the Fourier analysis of the function

$$\mathrm{f(t)} = \sin(\alpha_1 \cdot \mathrm{t}) \cdot \sin(\alpha_2 \cdot \mathrm{t}) \tag{3.8}$$

the coefficient $a_0 \neq 0$, which indicates the DC component, is obtained.

By way of explanation: While in the initial consideration the α_n were values dependent on t, i.e., included the variable t, in the function f(t) and in the following text, the α_n used do not include the variable t; therefore, it is appended multiplicatively.

The Fourier analysis describes the decomposition of a periodic function as follows:

$$\mathrm{f(t)} = \frac{a_0}{2} + \sum_{\mathrm{k}=1}^{\infty} (\mathrm{a_k} \cdot \cos(\mathrm{k} \cdot \mathrm{t}) + \mathrm{b_k} \cdot \sin(\mathrm{k} \cdot \mathrm{t})) \tag{3.9}$$

with

$$\mathrm{a_k} = \frac{1}{\pi} \int_{-\pi}^{\pi} \mathrm{f(t)} \cdot \cos(\mathrm{k} \cdot \mathrm{t})\mathrm{dt} \quad \text{for} \quad \mathrm{k} \geq 0 \tag{3.10}$$

$$\mathrm{b_k} = \frac{1}{\pi} \int_{-\pi}^{\pi} \mathrm{f(t)} \cdot \sin(\mathrm{k} \cdot \mathrm{t})\mathrm{dt} \quad \text{for} \quad \mathrm{k} \geq 1 \tag{3.11}$$

For the analysis of the DC component, it is essential to determine the coefficient a_k:

$$\mathrm{a_k} = \frac{1}{\pi}\int_{-\pi}^{\pi} \mathrm{f(t)\cdot\cos(k\cdot t)dt} \quad \text{for} \quad \mathrm{k}=0 \tag{3.12}$$

with

$$\mathrm{f(t)} = \sin(\alpha_1\cdot \mathrm{t})\cdot\sin(\alpha_2\cdot \mathrm{t}) \tag{3.13}$$

$$a_k = \frac{1}{\pi}\cdot\int_{-\pi}^{\pi}\sin(\alpha_1\cdot t)\cdot\sin(\alpha_2\cdot t)\cdot\cos(k\cdot t)dt$$

For k = 0, this yields

$$a_0 = \frac{1}{\pi}\cdot\int_{-\pi}^{\pi}\sin(\alpha_1\cdot t)\cdot\sin(\alpha_2\cdot t)dt \tag{3.14}$$

Auxiliary calculation (see also Bronstein [1]):

$$\int u'\cdot v dt = u\cdot v - \int u\cdot v'\, dt \tag{3.15}$$

$$\int u\prime\cdot v dt = u\cdot v - \left(u^+\cdot v\prime - \int u^+\cdot v'' dt\right) \tag{3.16}$$

$$\int u'\cdot v dt = u\cdot v - u^+\cdot v' + \int u^+\cdot v'' dt \tag{3.17}$$

$$\int u'\cdot v dt - \int u^+\cdot v'' dt = u\cdot v - u^+\cdot v' \tag{3.18}$$

End of auxiliary calculation

$$\begin{aligned}&\int_{-\pi}^{\pi}\sin(\alpha_1\cdot t)\cdot\sin(\alpha_2\cdot t)dt - \frac{\alpha_2^2}{\alpha_1^2}\cdot\int_{-\pi}^{\pi}\sin(\alpha_1\cdot t)\cdot\sin(\alpha_2\cdot t)dt\\&= -\frac{1}{\alpha_1}\cdot\cos(\alpha_1\cdot t)\cdot\sin(\alpha_2\cdot t) + \frac{\alpha_2}{\alpha_1^2}\cdot\sin(\alpha_1\cdot t)\cdot\cos(\alpha_2\cdot t)\end{aligned} \tag{3.19}$$

$$\begin{aligned}&\left(1-\frac{\alpha_2^2}{\alpha_1^2}\right)\cdot\int_{-\pi}^{\pi}\sin(\alpha_1\cdot t)\cdot\sin(\alpha_2\cdot t)dt\\&= \left[-\frac{1}{\alpha_1}\cdot\cos(\alpha_1\cdot t)\cdot\sin(\alpha_2\cdot t) + \frac{\alpha_2}{\alpha_1^2}\cdot\sin(\alpha_1\cdot t)\cdot\cos(\alpha_2\cdot t)\right]_{-\pi}^{\pi}\end{aligned} \tag{3.20}$$

$$\left(1-\frac{\alpha_2^2}{\alpha_1^2}\right)\cdot\int_{-\pi}^{\pi}\sin(\alpha_1\cdot t)\cdot\sin(\alpha_2\cdot t)dt$$
$$=\left[-\frac{1}{\alpha_1}\cdot\cos(\alpha_1\cdot\pi)\cdot\sin(\alpha_2\cdot\pi)+\frac{\alpha_2}{\alpha_1^2}\cdot\sin(\alpha_1\cdot\pi)\cdot\cos(\alpha_2\cdot\pi)\right]-$$
$$\left[-\frac{1}{\alpha_1}\cdot\cos(\alpha_1\cdot(-\pi))\cdot\sin(\alpha_2\cdot(-\pi))+\frac{\alpha_2}{\alpha_1^2}\cdot\sin(\alpha_1\cdot(-\pi))\cdot\cos(\alpha_2\cdot(-\pi))\right] \tag{3.21}$$

$$\left(1-\frac{\alpha_2^2}{\alpha_1^2}\right)\cdot\int_{-\pi}^{\pi}\sin(\alpha_1\cdot t)\cdot\sin(\alpha_2\cdot t)dt$$
$$=-\frac{1}{\alpha_1}\cdot\cos(\alpha_1\cdot\pi)\cdot\sin(\alpha_2\cdot\pi)+\frac{1}{\alpha_1}\cdot\cos(\alpha_1\cdot(-\pi))\cdot\sin(\alpha_2\cdot(-\pi))+ \tag{3.22}$$
$$+\frac{\alpha_2}{\alpha_1^2}\cdot\sin(\alpha_1\cdot\pi)\cdot\cos(\alpha_2\cdot\pi)-\frac{\alpha_2}{\alpha_1^2}\cdot\sin(\alpha_1\cdot(-\pi))\cdot\cos(\alpha_2\cdot(-\pi))$$

$$\frac{\alpha_1^2-\alpha_2^2}{\alpha_1^2}\cdot\int_{-\pi}^{\pi}\sin(\alpha_1\cdot t)\cdot\sin(\alpha_2\cdot t)dt$$
$$=-\frac{2}{\alpha_1}\cdot\cos(\alpha_1\cdot\pi)\cdot\sin(\alpha_2\cdot\pi)+\frac{2\cdot\alpha_2}{\alpha_1^2}\cdot\sin(\alpha_1\cdot\pi)\cdot\cos(\alpha_2\cdot\pi) \tag{3.23}$$

$$\int_{-\pi}^{\pi}\sin(\alpha_1\cdot t)\cdot\sin(\alpha_2\cdot t)dt=\frac{2\cdot\alpha_2}{\alpha_1^2-\alpha_2^2}\cdot\sin(\alpha_1\cdot\pi)\cdot\cos(\alpha_2\cdot\pi)$$
$$-\frac{2\cdot\alpha_1}{\alpha_1^2-\alpha_2^2}\cdot\cos(\alpha_1\cdot\pi)\cdot\sin(\alpha_2\cdot\pi) \tag{3.24}$$

substitute into

$$a_0=\frac{1}{\pi}\cdot\int_{-\pi}^{\pi}\sin(\alpha_1\cdot t)\cdot\sin(\alpha_2\cdot t)dt \tag{3.25}$$

$$a_0=\frac{1}{\pi}\cdot\left(\frac{2\cdot\alpha_2}{\alpha_1^2-\alpha_2^2}\cdot\sin(\alpha_1\cdot\pi)\cdot\cos(\alpha_2\cdot\pi)-\frac{2\cdot\alpha_1}{\alpha_1^2-\alpha_2^2}\cdot\cos(\alpha_1\cdot\pi)\cdot\sin(\alpha_2\cdot\pi)\right) \tag{3.26}$$

$$a_0=\frac{2}{\pi\cdot\left(\alpha_1^2-\alpha_2^2\right)}\cdot(\alpha_2\cdot\sin(\alpha_1\cdot\pi)\cdot\cos(\alpha_2\cdot\pi)-\alpha_1\cdot\cos(\alpha_1\cdot\pi)\cdot\sin(\alpha_2\cdot\pi)) \tag{3.27}$$

The following cases must be considered:

1. For $\alpha_1 = \alpha_2$, a fraction of the type "0/0" arises, which means a limit evaluation must be performed for $\alpha_1 \rightarrow \alpha_2$:

$$a_0 = \lim_{\alpha_1 \to \alpha_2} \frac{2}{\pi \cdot \left(\alpha_1^2 - \alpha_2^2\right)} \cdot (\alpha_2 \cdot \sin(\alpha_1 \cdot \pi) \cdot \cos(\alpha_2 \cdot \pi) - \alpha_1 \cdot \cos(\alpha_1 \cdot \pi) \cdot \sin(\alpha_2 \cdot \pi)) \tag{3.28}$$

Application of l'Hospital's rule, differentiation with respect to α_1:

$$a_0 = \lim_{\alpha_1 \to \alpha_2} \frac{2 \cdot \pi \cdot \alpha_2 \cdot \cos(\alpha_1 \cdot \pi) \cdot \cos(\alpha_2 \cdot \pi) - 2 \cdot \cos(\alpha_1 \cdot \pi) \cdot \sin(\alpha_2 \cdot \pi) + 2 \cdot \pi \cdot \alpha_1 \cdot \sin(\alpha_1 \cdot \pi) \cdot \sin(\alpha_2 \cdot \pi)}{2 \cdot \pi \cdot \alpha_1} \tag{3.29}$$

$$a_0 = \lim_{\alpha_1 \to \alpha_2} \frac{2 \cdot \pi \cdot \alpha_2 \cdot \cos(\alpha_1 \cdot \pi) \cdot \cos(\alpha_2 \cdot \pi) + 2 \cdot \pi \cdot \alpha_1 \cdot \sin(\alpha_1 \cdot \pi) \cdot \sin(\alpha_2 \cdot \pi) - 2 \cdot \cos(\alpha_1 \cdot \pi) \cdot \sin(\alpha_2 \cdot \pi)}{2 \cdot \pi \cdot \alpha_1} \tag{3.30}$$

$$a_0 = \frac{2 \cdot \pi \cdot \alpha_2 \cdot \cos(\alpha_2 \cdot \pi) \cdot \cos(\alpha_2 \cdot \pi) + 2 \cdot \pi \cdot \alpha_2 \cdot \sin(\alpha_2 \cdot \pi) \cdot \sin(\alpha_2 \cdot \pi) - 2 \cdot \cos(\alpha_2 \cdot \pi) \cdot \sin(\alpha_2 \cdot \pi)}{2 \cdot \pi \cdot \alpha_2} \tag{3.31}$$

$$a_0 = \frac{2 \cdot \pi \cdot \alpha_2 - \sin(2 \cdot \pi \cdot \alpha_2)}{2 \cdot \pi \cdot \alpha_2} \tag{3.32}$$

$$a_0 = 1 - \frac{\sin(2 \cdot \pi \cdot \alpha_2)}{2 \cdot \pi \cdot \alpha_2} \tag{3.33}$$

For $\alpha_1 = \alpha_2$ and $\alpha_2 \neq 0$ (which is the case, since alternating fields are assumed), there is a solution $\neq 0$ for a_0, i.e., when the frequencies of the two interacting fields are equal, there is a persistent interaction between the sources of the frequencies (Figs. 3.1, 3.2, and 3.3).

2. For $\alpha_1 \neq \alpha_2$, the following consideration applies: The resultant function from the multiplicative coupling of two oscillations becomes periodic only when the resultant period of the two oscillations is reached. The resultant period of the resulting oscillation is the period after which it returns to the same state as at the beginning of the period.
   If both periods are related to a reference period using rational factors, the resultant period is obtained by a reference factor to the reference period, which is the product of the two individual factors. The frequency of the resulting oscillation in multiplicative coupling is the sum of the two individual frequencies, with the envelope having a frequency equal to the reciprocal of the resultant period.
   To obtain the two individual oscillations as complete packets, the two periods of the oscillations are multiplied by the LCM (least common multiple) of the denominators of the two frequency expressions:

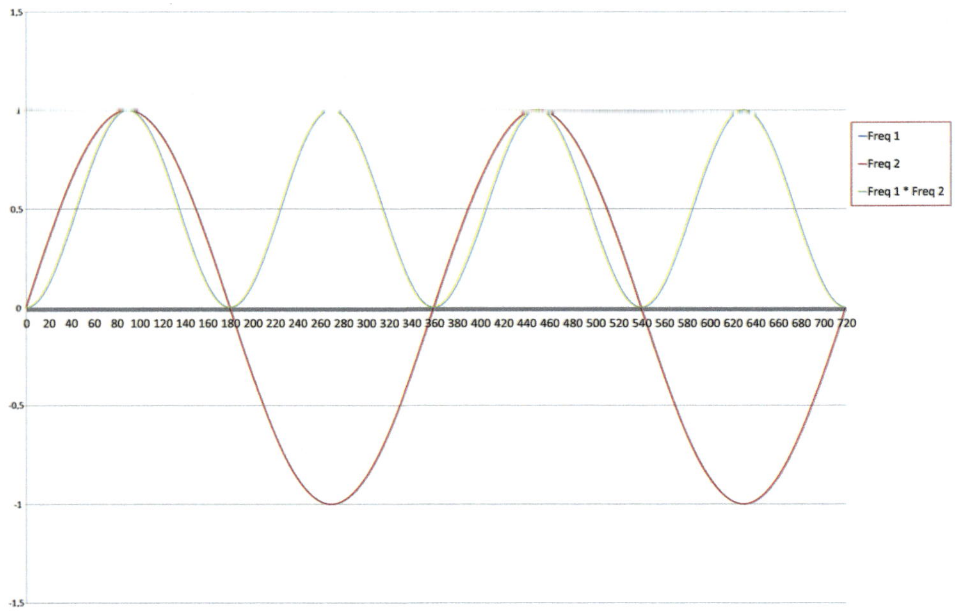

Fig. 3.1 Graph generated from an Excel spreadsheet illustrating the resultant effect between two alternating fields, Freq 1 (blue, behind the red line) and Freq 2 (red), with identical frequency, amplitude, and phase 0 between the two alternating fields; the resultant effect has a DC component, which in this case is positive.

> The envelopes of the two resulting oscillations have completed exactly one period after this time. Therefore, the mean value of these two resulting oscillations will always be zero.

This means that the two oscillations must have the same frequency if a persistent resultant force is to exist between the two interacting fields. Depending on the phase relationship between the two fields, the forces are either attractive or repulsive.

The more stable state of the two (repulsive or attractive) is when the two forces are attractive, because in attractive systems the distance decreases, if possible, and this reduction in distance simultaneously amplifies the force effect (distance appears in the denominator of the force calculation), causing the overall system to transition to a lower energy state. The amplifying effect of the mutual movement also has an adjusting effect on the phase relationship, so that the two fields attract each other even more strongly.

If the frequencies of the two fields are not equal and one is an irrational or inversely irrational multiple of the other, then the period duration of the resulting oscillation cannot be expressed in finite time. This means that, even in this case, a resultant force between the two fields could at least be detected for a while.

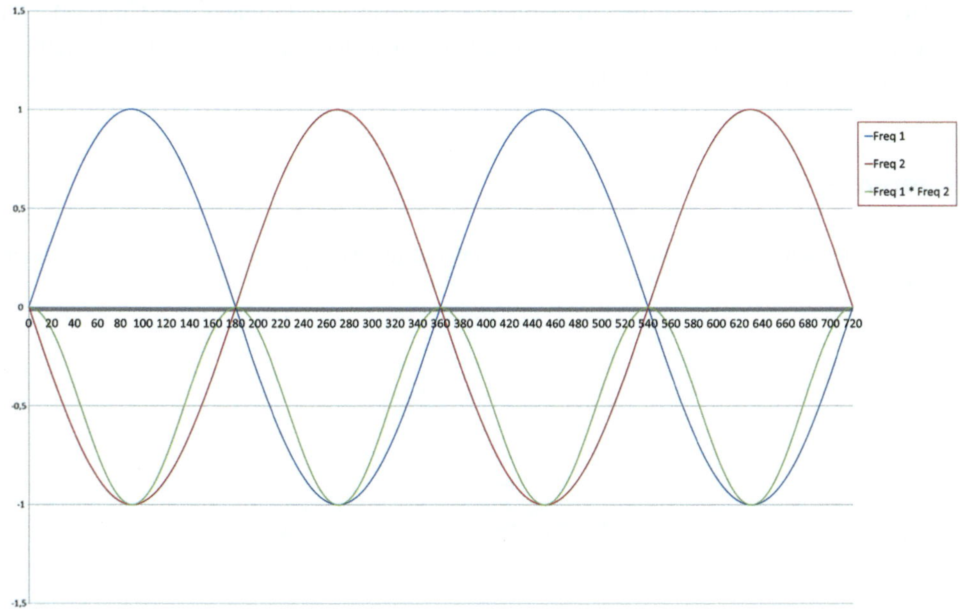

Fig. 3.2 Graph generated from an Excel spreadsheet illustrating the resultant effect between two alternating fields, Freq 1 (blue) and Freq 2 (red), with identical frequency and amplitude and a phase difference of 180° between the two alternating fields; the resultant effect has a DC component, which in this case is negative.

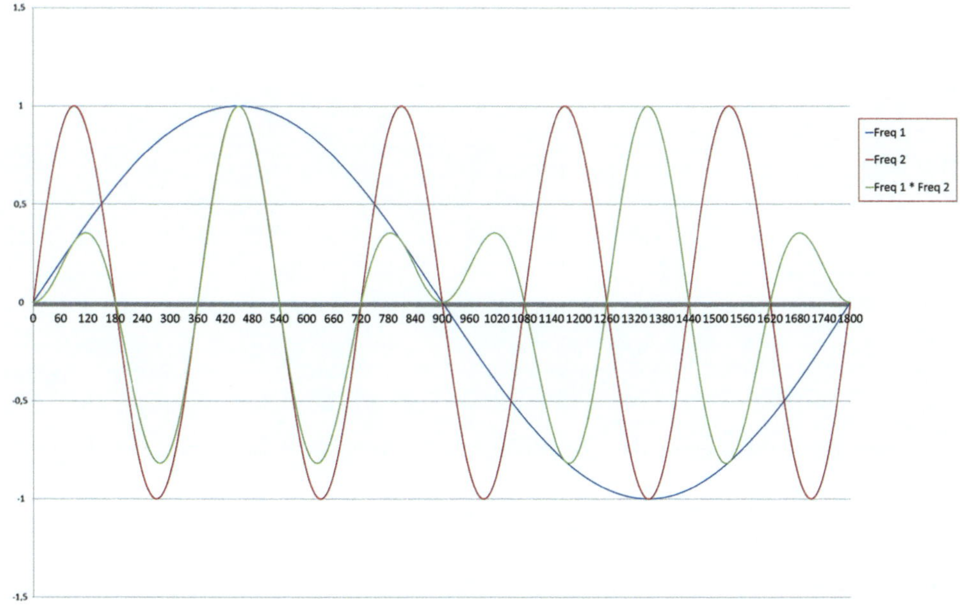

Fig. 3.3 Graph generated from an Excel spreadsheet illustrating the resultant effect between two alternating fields, Freq 1 (blue) and Freq 2 (red), with different frequencies (Freq 2 is 5×Freq 1), identical amplitude, and phase 0 between the two alternating fields; the resultant effect has no DC component and repeats periodically.

However, if it is assumed that the structure, however it is designed (see Chap. 4), is a generally applicable structure, it is rather unlikely that the frequencies of two structures have an irrational ratio to each other.

Reference

1. Bronstein, Semendjajew; Taschenbuch der Mathematik; Verlag Nauka Moskau, 25. Auflage 1991

The Toroidal Coil 4

Based on the preceding considerations, a spatially confined structure is sought that is capable of combining multiple oscillations within itself and, at the same time, maintaining these oscillations indefinitely. The resulting structure is a toroidal helix with properties derived below. These oscillations, which the toroidal helix unites, are later the causes of various physical effects such as gravitation, electric and magnetic fields, and so on.

Toroidal structures also appear in string theories [1; 2].

The so-called Maxwell's equations describe a system of electric and magnetic fields with certain interactions and properties (see Sect. 4.1):

- The vector describing an electric field and the vector describing the corresponding magnetic field, and vice versa, are always perpendicular to each other.
- A vector describing an electric or magnetic field generates a circular field whose plane is perpendicular to the vector.
- A vector describing an electric field that changes over time induces a vector describing a corresponding magnetic field with analogous temporal variations.

If one derives the lines of the hodographs from the respective vector fields of the electric and magnetic alternating fields, these lines are referred to as the field lines of the corresponding fields.

A spatially closed system in which there is a mutual interaction between an electric and a magnetic alternating field is based on a graphical system in which the field lines of the electric and corresponding magnetic fields each form a helix with a 45-degree pitch and a phase shift of 180° between the electric and magnetic fields (Figs. 4.1 and 4.2).

This system is described by the following equations.

J. von Stackelberg, *The Mass of a Photon*, https://doi.org/10.1007/978-3-658-49408-7_4

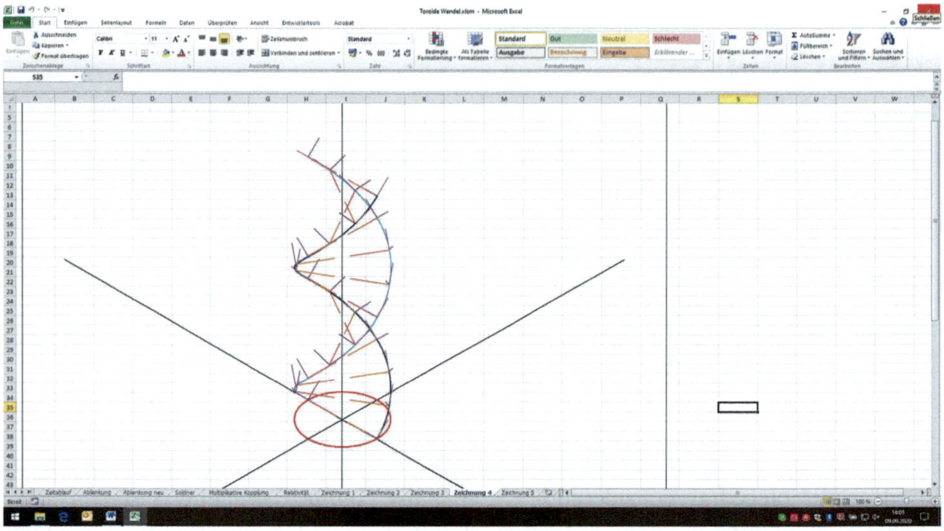

Fig. 4.1 Isometric view of two interwoven helices with the base circle (lying in the plane spanned by the x and y coordinate axes); at regular intervals along both helices, the associated tangent vectors (position vectors of the field lines), referenced to their respective points of origin, as well as the normal and binormal vectors, are shown

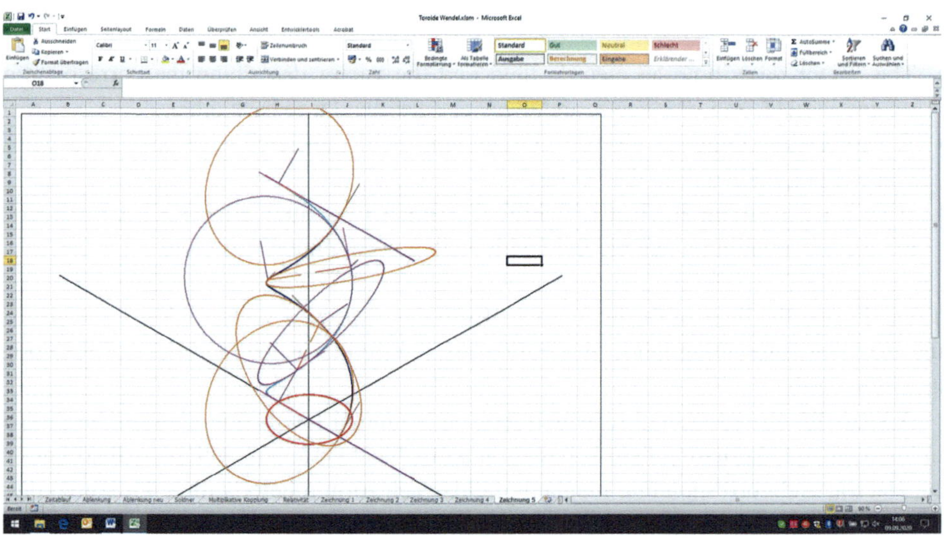

Fig. 4.2 Isometric view of two interwoven helices; if a circle with a radius equal to the distance between the two helices is placed around the tangent vector of one helix, the tangent vector of the other helix precisely forms the tangent vector for the circle, just as the relationship between electric and magnetic fields is described by Maxwell's equations

The position vector of the field line of the electric field is given by:

$$\overrightarrow{W_e} = \begin{pmatrix} x_e \\ y_e \\ z_e \end{pmatrix} = \begin{pmatrix} r \cdot \cos\alpha \\ r \cdot \sin\alpha \\ r \cdot \alpha \end{pmatrix} \tag{4.1}$$

The position vector of the field line of the magnetic field includes the phase position with a 180° shift on the base circle:

$$\overrightarrow{W_m} = \begin{pmatrix} x_m \\ y_m \\ z_m \end{pmatrix} = \begin{pmatrix} r \cdot \cos(\alpha + 180) \\ r \cdot \sin(\alpha + 180) \\ r \cdot \alpha \end{pmatrix} \tag{4.2}$$

$$\overrightarrow{W_m} = \begin{pmatrix} x_m \\ y_m \\ z_m \end{pmatrix} = \begin{pmatrix} -r \cdot \cos\alpha \\ -r \cdot \sin\alpha \\ r \cdot \alpha \end{pmatrix} \tag{4.3}$$

These two helices form the hodographs of the vector fields for the electric and magnetic fields. The vectors are ultimately the tangent vectors of the helices.

The tangent vectors can be calculated by differentiating with respect to α:

$$\overrightarrow{T_e} = \frac{d}{d\alpha} \begin{pmatrix} x_e \\ y_e \\ z_e \end{pmatrix} = \frac{d}{d\alpha} \begin{pmatrix} r \cdot \cos\alpha \\ r \cdot \sin\alpha \\ r \cdot \alpha \end{pmatrix} = r \cdot \begin{pmatrix} -\sin\alpha \\ \cos\alpha \\ 1 \end{pmatrix} \tag{4.4}$$

$$\overrightarrow{T_m} = \frac{d}{d\alpha} \begin{pmatrix} x_m \\ y_m \\ z_m \end{pmatrix} = \frac{d}{d\alpha} \begin{pmatrix} -r \cdot \cos\alpha \\ -r \cdot \sin\alpha \\ r \cdot \alpha \end{pmatrix} = r \cdot \begin{pmatrix} \sin\alpha \\ -\cos\alpha \\ 1 \end{pmatrix} \tag{4.5}$$

To calculate the accompanying triad of the helices, the two normal vectors must also be determined. The normal vectors can be calculated from the position vector of the points on the helix and the z-coordinate linked to the respective angle α:

$$\vec{N} = \vec{Z} - \vec{W} \tag{4.6}$$

$$\overrightarrow{N_e} = \begin{pmatrix} 0 \\ 0 \\ r \cdot \alpha \end{pmatrix} - \begin{pmatrix} r \cdot \cos\alpha \\ r \cdot \sin\alpha \\ r \cdot \alpha \end{pmatrix} \tag{4.7}$$

$$\overrightarrow{N_e} = \begin{pmatrix} -r \cdot \cos\alpha \\ -r \cdot \sin\alpha \\ 0 \end{pmatrix} \tag{4.8}$$

$$\overrightarrow{N_m} = \begin{pmatrix} 0 \\ 0 \\ r \cdot \alpha \end{pmatrix} - \begin{pmatrix} -r \cdot \cos\alpha \\ -r \cdot \sin\alpha \\ r \cdot \alpha \end{pmatrix} \tag{4.9}$$

$$\overrightarrow{N_m} = \begin{pmatrix} r \cdot \cos\alpha \\ r \cdot \sin\alpha \\ 0 \end{pmatrix} \tag{4.10}$$

From the tangent and normal vectors, the binormal vectors can be calculated using the cross product:

$$\vec{B} = \vec{T} \times \vec{N} \tag{4.11}$$

$$\overrightarrow{B_e} = \overrightarrow{T_e} \times \overrightarrow{N_e} \tag{4.12}$$

$$\overrightarrow{B_e} = r \cdot \begin{pmatrix} -\sin\alpha \\ \cos\alpha \\ 1 \end{pmatrix} \times \begin{pmatrix} -r \cdot \cos\alpha \\ -r \cdot \sin\alpha \\ 0 \end{pmatrix} \tag{4.13}$$

$$\overrightarrow{B_e} = r^2 \cdot \begin{pmatrix} -\sin\alpha \\ \cos\alpha \\ 1 \end{pmatrix} \times \begin{pmatrix} -\cos\alpha \\ -\sin\alpha \\ 0 \end{pmatrix} \tag{4.14}$$

$$\overrightarrow{B_e} = r^2 \cdot \begin{pmatrix} \sin\alpha \\ -\cos\alpha \\ 1 \end{pmatrix} \tag{4.15}$$

$$\overrightarrow{B_m} = \overrightarrow{T_m} \times \overrightarrow{N_m} \tag{4.16}$$

$$\overrightarrow{B_m} = r \cdot \begin{pmatrix} \sin\alpha \\ -\cos\alpha \\ 1 \end{pmatrix} \times \begin{pmatrix} r \cdot \cos\alpha \\ r \cdot \sin\alpha \\ 0 \end{pmatrix} \tag{4.17}$$

$$\overrightarrow{B_m} = r^2 \cdot \begin{pmatrix} \sin\alpha \\ -\cos\alpha \\ 1 \end{pmatrix} \times \begin{pmatrix} \cos\alpha \\ \sin\alpha \\ 0 \end{pmatrix} \tag{4.18}$$

$$\overrightarrow{B_m} = r^2 \cdot \begin{pmatrix} -\sin\alpha \\ \cos\alpha \\ 1 \end{pmatrix} \tag{4.19}$$

The interrelations between the electric and magnetic fields are expressed in Maxwell's equations in their differential form (see also Sect. 4.1.3) as follows:

$$rot\,\vec{E} = -\frac{d\vec{B}}{dt} \tag{4.20}$$

$$rot\,\vec{H} = \vec{j_l} + \frac{d\vec{D}}{dt} \tag{4.21}$$

with

- E = electric field strength
- B = magnetic flux density
- H = magnetic field strength
- D = electric displacement density
- J_l = electric current
- t = time

As shown in Sect. 4.1.5, the rotation operation does not generally work in three-dimensional space. Therefore, the consideration of whether the two toroidal helices correspond to the structure of electric and magnetic fields must be carried out step by step.

The tangent vector of one helix and the binormal vector of the second helix are related in the same way as electric and magnetic fields: one forms concentric circles around the other, with the normal vector of the concentric circles represented by the binormal vector of the opposing helix, which is valid only for a single point on the helix.

For this purpose, the tangent vector of one helix and the binormal vector of the other helix must be parallel:

$$\overrightarrow{T_e} = r \cdot \begin{pmatrix} -\sin\alpha \\ \cos\alpha \\ 1 \end{pmatrix} \tag{4.22}$$

$$\overrightarrow{B_m} = r^2 \cdot \begin{pmatrix} -\sin\alpha \\ \cos\alpha \\ 1 \end{pmatrix} \tag{4.23}$$

$$\overrightarrow{T_m} = r \cdot \begin{pmatrix} \sin\alpha \\ -\cos\alpha \\ 1 \end{pmatrix} \tag{4.24}$$

$$\overrightarrow{B_e} = r^2 \cdot \begin{pmatrix} \sin\alpha \\ -\cos\alpha \\ 1 \end{pmatrix} \tag{4.25}$$

It is easy to see that, apart from their magnitudes, the two vectors are parallel. It follows that the magnetic and electric fields complement each other when arranged in the form of the two helices described above.

To clarify the relationship, the two position vectors for the enclosing circular shapes of the two helices are determined.

Constructing the circle at a point on the helix with the tangent vector of the helix as the normal vector is achieved using the parametric form of the plane on which the circle lies, with the normal and binormal vectors of the accompanying trihedron spanning the plane.

1. The position vector of the helix point serves as the reference point of the plane on which the circle lies:

$$\overrightarrow{W_e} = \begin{pmatrix} r\cdot\cos\alpha \\ r\cdot\sin\alpha \\ r\cdot\alpha \end{pmatrix} \tag{4.26}$$

2. The tangent vector at the helix point is the normal vector for the plane on which the circle lies:

$$\overrightarrow{T_e} = \begin{pmatrix} -r\cdot\sin\alpha \\ r\cdot\cos\alpha \\ r \end{pmatrix} \tag{4.27}$$

3. The normal vector and the binormal vector span the plane on which the circle lies:

$$\overrightarrow{N_e} = \begin{pmatrix} -r\cdot\cos\alpha \\ -r\cdot\sin\alpha \\ 0 \end{pmatrix}, \overrightarrow{N_{en}} = \begin{pmatrix} -\cos\alpha \\ -\sin\alpha \\ 0 \end{pmatrix}, \tag{4.28}$$

$$\overrightarrow{B_e} = \begin{pmatrix} r^2\cdot\sin\alpha \\ -r^2\cdot\cos\alpha \\ r^2 \end{pmatrix}, \overrightarrow{B_{en}} = \begin{pmatrix} \sin\alpha \\ -\cos\alpha \\ 1 \end{pmatrix}, \tag{4.29}$$

4. From the position vector of the reference point, the normal vector, and the binormal vector, the parametric form of the plane can be constructed, or the circle can be represented by choosing sine and cosine of the parameter as the two variables. The magnitude of the sum of the two vectors, each multiplied by its respective parameter, equals the radius of the circle. The normal and binormal vectors are orthogonal to each other; therefore, they must be multiplied by the cosine (normal vector) and sine (binormal vector) of the parameter φ for the circle:

The parametric form of the plane:

$$\overrightarrow{K_m} = \overrightarrow{W_e} + s \cdot \overrightarrow{N_{en}} + t \cdot \overrightarrow{B_{en}} \tag{4.30}$$

The length condition for the circle radius:

$$\left| s \cdot \overrightarrow{N_{en}} + t \cdot \overrightarrow{B_{en}} \right| = r_{Kreis} \tag{4.31}$$

$$\text{I} \quad \overrightarrow{K_m} = \overrightarrow{W_e} + s \cdot \cos\varphi \cdot \overrightarrow{N_{en}} + t \cdot \sin\varphi \cdot \overrightarrow{B_{en}} \tag{4.32}$$

$$\text{II} \quad s \cdot \left| \overrightarrow{N_{en}} \right| = r_{Kreis} \tag{4.33}$$

$$\text{III} \quad t \cdot \left| \overrightarrow{B_{en}} \right| = r_{Kreis} \tag{4.34}$$

$$\text{II} \quad s \cdot \left| \begin{pmatrix} -\cos\alpha \\ -\sin\alpha \\ 0 \end{pmatrix} \right| = 2 \cdot r \tag{4.35}$$

$$\text{III} \quad t \cdot \left| \begin{pmatrix} \sin\alpha \\ -\cos\alpha \\ 1 \end{pmatrix} \right| = 2 \cdot r \tag{4.36}$$

$$\text{II} \quad s = 2 \cdot r \text{ in I} \tag{4.37}$$

$$\text{III} \quad t \cdot \sqrt{1+1} = 2 \cdot r \tag{4.38}$$

$$\text{III} \quad t = \sqrt{\frac{4 \cdot r^2}{2}}$$

$$\text{III} \quad t = \sqrt{2} \cdot r \text{ in I} \tag{4.39}$$

$$\mathrm{I}\quad \overrightarrow{\mathrm{K_m}} = \begin{pmatrix} \mathrm{r}\cdot\cos\alpha \\ \mathrm{r}\cdot\sin\alpha \\ \mathrm{r}\cdot\alpha \end{pmatrix} + 2\cdot\mathrm{r}\cdot\cos\varphi\cdot\overrightarrow{\mathrm{N_{en}}} + \sqrt{2}\cdot\mathrm{r}\cdot\sin\varphi\cdot\overrightarrow{\mathrm{B_{en}}} \tag{4.40}$$

$$\mathrm{I}\quad \overrightarrow{\mathrm{K_m}} = \mathrm{r}\cdot\begin{pmatrix} \cos\alpha \\ \sin\alpha \\ \alpha \end{pmatrix} + 2\cdot\mathrm{r}\cdot\cos\varphi\cdot\begin{pmatrix} -\cos\alpha \\ -\sin\alpha \\ 0 \end{pmatrix} + \sqrt{2}\cdot\mathrm{r}\cdot\sin\varphi\cdot\begin{pmatrix} \sin\alpha \\ -\cos\alpha \\ 1 \end{pmatrix} \tag{4.41}$$

A corresponding position vector for the circular points of the electric field line is obtained analogously:

$$\overrightarrow{\mathrm{W_m}} = \begin{pmatrix} -\mathrm{r}\cdot\cos\alpha \\ -\mathrm{r}\cdot\sin\alpha \\ \mathrm{r}\cdot\alpha \end{pmatrix} \tag{4.42}$$

$$\overrightarrow{\mathrm{T_m}} = \mathrm{r}\cdot\begin{pmatrix} \sin\alpha \\ -\cos\alpha \\ 1 \end{pmatrix} \tag{4.43}$$

$$\overrightarrow{\mathrm{N_m}} = \begin{pmatrix} \mathrm{r}\cdot\cos\alpha \\ \mathrm{r}\cdot\sin\alpha \\ 0 \end{pmatrix}, \quad \overrightarrow{\mathrm{N_{mn}}} = \begin{pmatrix} \cos\alpha \\ \sin\alpha \\ 0 \end{pmatrix} \tag{4.44}$$

$$\overrightarrow{\mathrm{B_m}} = \mathrm{r}^2\cdot\begin{pmatrix} -\sin\alpha \\ \cos\alpha \\ 1 \end{pmatrix}, \overrightarrow{\mathrm{B_{mn}}} = \begin{pmatrix} -\sin\alpha \\ \cos\alpha \\ 1 \end{pmatrix} \tag{4.45}$$

$$\overrightarrow{\mathrm{K_e}} = \mathrm{r}\cdot\begin{pmatrix} -\cos\alpha \\ -\sin\alpha \\ \alpha \end{pmatrix} + 2\cdot\mathrm{r}\cdot\cos\varphi\cdot\begin{pmatrix} \cos\alpha \\ \sin\alpha \\ 0 \end{pmatrix} + \sqrt{2}\cdot\mathrm{r}\cdot\sin\varphi\cdot\begin{pmatrix} -\sin\alpha \\ \cos\alpha \\ 1 \end{pmatrix} \tag{4.46}$$

Further details on the temporal behavior of the two helices relative to each other are described in 4.1 in the section on Maxwell's equations.

The cylinder enclosed by the two graphs must now be bent into a toroid in order to transform a one-dimensional infinitely long system into a spatially closed system (Fig. 4.3).

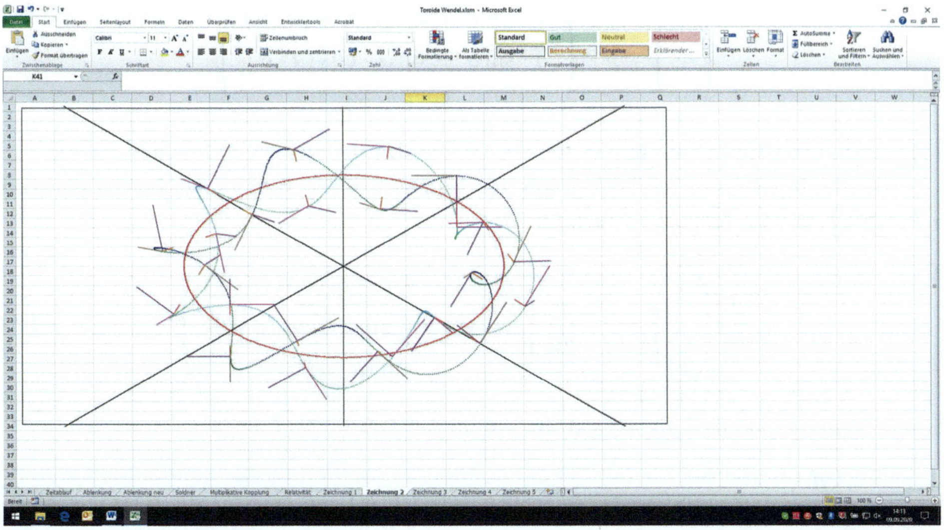

Fig. 4.3 Isometric representation of the intertwined helices bent into a toroid, with indicated tangent, normal, and binormal vectors

The position vector of the points of the helix graph around the toroid can be described as follows:

$$\begin{pmatrix} x \\ y \\ z \end{pmatrix} = \begin{pmatrix} R \cdot \cos\alpha + r \cdot \cos\alpha \cdot \cos\beta \\ R \cdot \sin\alpha + r \cdot \sin\alpha \cdot \cos\beta \\ r \cdot \sin\beta \end{pmatrix} \tag{4.47}$$

with

R = radius of the carrier circle for the toroid
r = radius of the toroid
α = angle for the carrier circle
β = angle for the toroidal circle

To produce an average slope of 45°, the ratio of the two radii and the ratio of the development of the two angles α and β must be identical:

$$\frac{R}{r} = \frac{\beta}{\alpha} = a \tag{4.48}$$

This yields

$$\begin{pmatrix} x \\ y \\ z \end{pmatrix} = \begin{pmatrix} a \cdot r \cdot \cos\alpha + r \cdot \cos\alpha \cdot \cos(a \cdot \alpha) \\ a \cdot r \cdot \sin\alpha + r \cdot \sin\alpha \cdot \cos(a \cdot \alpha) \\ r \cdot \sin(a \cdot \alpha) \end{pmatrix} \tag{4.49}$$

$$\begin{pmatrix} x \\ y \\ z \end{pmatrix} = r \cdot \begin{pmatrix} a \cdot \cos\alpha + \cos\alpha \cdot \cos(a \cdot \alpha) \\ a \cdot \sin\alpha + \sin\alpha \cdot \cos(a \cdot \alpha) \\ \sin(a \cdot \alpha) \end{pmatrix} \tag{4.50}$$

This helix forms the hodograph of the vector field for the electric or magnetic field.

From the function of the hodograph, the function of the vectors can be calculated by differentiation with respect to the angle α.

$$\frac{d}{d\alpha}\begin{pmatrix} x \\ y \\ z \end{pmatrix} = r \cdot \frac{d}{d\alpha}\begin{pmatrix} a \cdot \cos\alpha + \cos\alpha \cdot \cos(a \cdot \alpha) \\ a \cdot \sin\alpha + \sin\alpha \cdot \cos(a \cdot \alpha) \\ \sin(a \cdot \alpha) \end{pmatrix} \tag{4.51}$$

Supplementary calculation (see also Bronstein):

$$\frac{d}{d\alpha}(\cos\alpha \cdot \cos(a \cdot \alpha)) = -\sin\alpha \cdot \cos(a \cdot \alpha) - a \cdot \cos\alpha \cdot \sin(a \cdot \alpha) \tag{4.52}$$

$$\frac{d}{d\alpha}(\sin\alpha \cdot \cos(a \cdot \alpha)) = \cos\alpha \cdot \cos(a \cdot \alpha) - a \cdot \sin\alpha \cdot \sin(a \cdot \alpha) \tag{4.53}$$

End of supplementary calculation

$$\frac{d}{d\alpha}\begin{pmatrix} x \\ y \\ z \end{pmatrix} = r \cdot \begin{pmatrix} -a \cdot \sin\alpha - \sin\alpha \cdot \cos(a \cdot \alpha) - a \cdot \cos\alpha \cdot \sin(a \cdot \alpha) \\ a \cdot \cos\alpha + \cos\alpha \cdot \cos(a \cdot \alpha) - a \cdot \sin\alpha \cdot \sin(a \cdot \alpha) \\ a \cdot \cos(a \cdot \alpha) \end{pmatrix} \tag{4.54}$$

This vector is simultaneously the tangent vector T.

$$\vec{T} = r \cdot \begin{pmatrix} -a \cdot \sin\alpha - \sin\alpha \cdot \cos(a \cdot \alpha) - a \cdot \cos\alpha \cdot \sin(a \cdot \alpha) \\ a \cdot \cos\alpha + \cos\alpha \cdot \cos(a \cdot \alpha) - a \cdot \sin\alpha \cdot \sin(a \cdot \alpha) \\ a \cdot \cos(a \cdot \alpha) \end{pmatrix} \tag{4.55}$$

To determine the accompanying triad, the normal vector N must also be calculated.

In the case of the toroidal helix, the normal vector can be determined as the difference between the position vector of the base circle K and the position vector of the helix W for a given angle α.

$$\vec{K} = \begin{pmatrix} a \cdot r \cdot \cos\alpha \\ a \cdot r \cdot \sin\alpha \\ 0 \end{pmatrix} \tag{4.56}$$

$$\vec{W} = r \cdot \begin{pmatrix} a \cdot \cos\alpha + \cos\alpha \cdot \cos(a \cdot \alpha) \\ a \cdot \sin\alpha + \sin\alpha \cdot \cos(a \cdot \alpha) \\ \sin(a \cdot \alpha) \end{pmatrix} \tag{4.57}$$

$$\vec{N} = \vec{K} - \vec{W} \tag{4.58}$$

$$\vec{N} = \begin{pmatrix} a \cdot r \cdot \cos\alpha \\ a \cdot r \cdot \sin\alpha \\ 0 \end{pmatrix} - r \cdot \begin{pmatrix} a \cdot \cos\alpha + \cos\alpha \cdot \cos(a \cdot \alpha) \\ a \cdot \sin\alpha + \sin\alpha \cdot \cos(a \cdot \alpha) \\ \sin(a \cdot \alpha) \end{pmatrix} \tag{4.59}$$

$$\vec{N} = r \cdot \begin{pmatrix} a \cdot \cos\alpha - a \cdot \cos\alpha - \cos\alpha \cdot \cos(a \cdot \alpha) \\ a \cdot \sin\alpha - a \cdot \sin\alpha - \sin\alpha \cdot \cos(a \cdot \alpha) \\ -\sin(a \cdot \alpha) \end{pmatrix} \tag{4.60}$$

$$\vec{N} = r \cdot \begin{pmatrix} -\cos\alpha \cdot \cos(a \cdot \alpha) \\ -\sin\alpha \cdot \cos(a \cdot \alpha) \\ -\sin(a \cdot \alpha) \end{pmatrix} \tag{4.61}$$

From the tangent vector T and the normal vector N, the binormal vector B can be calculated as the cross product of these two vectors.

$$\vec{B} = \vec{T} \times \vec{N} \tag{4.62}$$

$$\vec{B} = r \cdot \begin{pmatrix} -a \cdot \sin\alpha - \sin\alpha \cdot \cos(a \cdot \alpha) - a \cdot \cos\alpha \cdot \sin(a \cdot \alpha) \\ a \cdot \cos\alpha + \cos\alpha \cdot \cos(a \cdot \alpha) - a \cdot \sin\alpha \cdot \sin(a \cdot \alpha) \\ a \cdot \cos(a \cdot \alpha) \end{pmatrix} \times r \cdot \begin{pmatrix} -\cos\alpha \cdot \cos(a \cdot \alpha) \\ -\sin\alpha \cdot \cos(a \cdot \alpha) \\ -\sin(a \cdot \alpha) \end{pmatrix} \tag{4.63}$$

$$\vec{B} = r^2 \cdot \begin{pmatrix} -a \cdot \sin\alpha - \sin\alpha \cdot \cos(a \cdot \alpha) - a \cdot \cos\alpha \cdot \sin(a \cdot \alpha) \\ a \cdot \cos\alpha + \cos\alpha \cdot \cos(a \cdot \alpha) - a \cdot \sin\alpha \cdot \sin(a \cdot \alpha) \\ a \cdot \cos(a \cdot \alpha) \end{pmatrix} \times \begin{pmatrix} -\cos\alpha \cdot \cos(a \cdot \alpha) \\ -\sin\alpha \cdot \cos(a \cdot \alpha) \\ -\sin(a \cdot \alpha) \end{pmatrix} \tag{4.64}$$

$$\vec{B} = r^2 \cdot \begin{pmatrix} (a \cdot \cos\alpha + \cos\alpha \cdot \cos(a \cdot \alpha) - a \cdot \sin\alpha \cdot \sin(a \cdot \alpha)) \cdot (-\sin(a \cdot \alpha)) - \\ -(-\sin\alpha \cdot \cos(a \cdot \alpha)) \cdot (a \cdot \cos(a \cdot \alpha)) \\ (a \cdot \cos(a \cdot \alpha)) \cdot (-\cos\alpha \cdot \cos(a \cdot \alpha)) - \\ -(-\sin(a \cdot \alpha)) \cdot (-a \cdot \sin\alpha - \sin\alpha \cdot \cos(a \cdot \alpha) - a \cdot \cos\alpha \cdot \sin(a \cdot \alpha)) \\ (-a \cdot \sin\alpha - \sin\alpha \cdot \cos(a \cdot \alpha) - a \cdot \cos\alpha \cdot \sin(a \cdot \alpha)) \cdot (-\sin\alpha \cdot \cos(a \cdot \alpha)) - \\ -(-\cos\alpha \cdot \cos(a \cdot \alpha)) \cdot (a \cdot \cos\alpha + \cos\alpha \cdot \cos(a \cdot \alpha) - a \cdot \sin\alpha \cdot \sin(a \cdot \alpha)) \end{pmatrix} \tag{4.65}$$

$$\vec{B}=r^2\cdot\left(\begin{array}{l}(a\cdot\cos\alpha\cdot(-\sin(a\cdot\alpha))+\cos\alpha\cdot\cos(a\cdot\alpha)\cdot(-\sin(a\cdot\alpha))-a\cdot\sin\alpha\cdot\sin(a\cdot\alpha)\cdot(-\sin(a\cdot\alpha)))-\\ -(-\sin\alpha\cdot\cos(a\cdot\alpha))\cdot(a\cdot\cos(a\cdot\alpha))\\ (-(-a\cdot\cos(a))\cdot(-\cos\alpha\cdot\cos(a\cdot\alpha))-\\ -(-\sin(a\cdot\alpha))-\sin\alpha\cdot\cos(a\cdot\alpha)\cdot(-\sin(a\cdot\alpha))-a\cdot\cos\alpha\cdot\sin(a\cdot\alpha)\cdot(-\sin(a\cdot\alpha)))\\ \qquad\left(\begin{array}{l}-a\cdot\sin\alpha\cdot(-\sin\alpha\cdot\cos(a\cdot\alpha))-\\ -\sin\alpha\cdot\cos(a\cdot\alpha)\cdot(-\sin\alpha\cdot\cos(a\cdot\alpha))-\\ -a\cdot\cos\alpha\cdot\sin(a\cdot\alpha)\cdot(-\sin\alpha\cdot\cos(a\cdot\alpha))\end{array}\right)+ \qquad +1\\ \qquad+\left(\begin{array}{l}a\cdot\cos\alpha\cdot\cos(a\cdot\alpha)\cdot\cos\alpha+\cos\alpha\cdot\cos(a\cdot\alpha)\cdot\cos\alpha\cdot\cos(a\cdot\alpha)-\\ -a\cdot\cos\alpha\cdot\cos(a\cdot\alpha)\cdot\sin\alpha\cdot\sin(a\cdot\alpha)\end{array}\right)\end{array}\right) \tag{4.66}$$

$$\overrightarrow{B}=r^2\cdot\begin{pmatrix}-\cos\alpha\cdot\sin(a\cdot\alpha)\cdot\cos(a\cdot\alpha)+a\cdot\sin\alpha-a\cdot\cos\alpha\cdot\sin(a\cdot\alpha)\\ -\sin\alpha\cdot\sin(a\cdot\alpha)\cdot\cos(a\cdot\alpha)-a\cdot\cos\alpha-a\cdot\sin\alpha\cdot\sin(a\cdot\alpha)\\ \cos(a\cdot\alpha)\cdot\cos(a\cdot\alpha)+a\cdot\cos(a\cdot\alpha)\end{pmatrix} \tag{4.67}$$

A second helix winds around the toroid at a phase angle of 180°, relative to the winding. If the first helix is described as follows:

$$\overrightarrow{W_1}=r\cdot\begin{pmatrix}a\cdot\cos\alpha+\cos\alpha\cdot\cos(a\cdot\alpha)\\ a\cdot\sin\alpha+\sin\alpha\cdot\cos(a\cdot\alpha)\\ \sin(a\cdot\alpha)\end{pmatrix}, \tag{4.68}$$

then the second helix is described as follows:

$$\overrightarrow{W_2}=r\cdot\begin{pmatrix}a\cdot\cos\alpha+\cos\alpha\cdot\cos(a\cdot\alpha+\pi)\\ a\cdot\sin\alpha+\sin\alpha\cdot\cos(a\cdot\alpha+\pi)\\ \sin(a\cdot\alpha+\pi)\end{pmatrix} \tag{4.69}$$

$$\overrightarrow{W_2}=r\cdot\begin{pmatrix}a\cdot\cos\alpha+\cos\alpha\cdot(\cos(a\cdot\alpha)\cdot\cos\pi-\sin(a\cdot\alpha)\cdot\sin\pi)\\ a\cdot\sin\alpha+\sin\alpha\cdot(\cos(a\cdot\alpha)\cdot\cos\pi-\sin(a\cdot\alpha)\cdot\sin\pi)\\ (\sin(a\cdot\alpha)\cdot\cos\pi+\cos(a\cdot\alpha)\cdot\sin\pi)\end{pmatrix} \tag{4.70}$$

$$\overrightarrow{W_2}=r\cdot\begin{pmatrix}a\cdot\cos\alpha-\cos\alpha\cdot\cos(a\cdot\alpha)\\ a\cdot\sin\alpha-\sin\alpha\cdot\cos(a\cdot\alpha)\\ -\sin(a\cdot\alpha)\end{pmatrix} \tag{4.71}$$

From this helical function, the tangent can be calculated by differentiating with respect to α:

$$\overrightarrow{T_2} = r \cdot \frac{d}{d\alpha} \begin{pmatrix} a \cdot \cos\alpha - \cos\alpha \cdot \cos(a \cdot \alpha) \\ a \cdot \sin\alpha - \sin\alpha \cdot \cos(a \cdot \alpha) \\ -\sin(a \cdot \alpha) \end{pmatrix} \tag{4.72}$$

$$\overrightarrow{T_2} = r \cdot \begin{pmatrix} -a \cdot \sin\alpha + \sin\alpha \cdot \cos(a \cdot \alpha) + a \cdot \cos\alpha \cdot \sin(a \cdot \alpha) \\ a \cdot \cos\alpha - \cos\alpha \cdot \cos(a \cdot \alpha) + a \cdot \sin\alpha \cdot \sin(a \cdot \alpha) \\ -a \cdot \cos(a \cdot \alpha) \end{pmatrix} \tag{4.73}$$

The normal vector for the second helix can be calculated analogously to that of the first helix:

$$\overrightarrow{N_2} = \overrightarrow{K} - \overrightarrow{W_2} \tag{4.74}$$

$$\overrightarrow{N_2} = r \cdot \begin{pmatrix} a \cdot \cos\alpha \\ a \cdot \sin\alpha \\ 0 \end{pmatrix} - r \cdot \begin{pmatrix} a \cdot \cos\alpha - \cos\alpha \cdot \cos(a \cdot \alpha) \\ a \cdot \sin\alpha - \sin\alpha \cdot \cos(a \cdot \alpha) \\ -\sin(a \cdot \alpha) \end{pmatrix} \tag{4.75}$$

$$\overrightarrow{N_2} = r \cdot \begin{pmatrix} a \cdot \cos\alpha - a \cdot \cos\alpha + \cos\alpha \cdot \cos(a \cdot \alpha) \\ a \cdot \sin\alpha - a \cdot \sin\alpha + \sin\alpha \cdot \cos(a \cdot \alpha) \\ \sin(a \cdot \alpha) \end{pmatrix} \tag{4.76}$$

The binormal vector is calculated from the tangent vector and the normal vector.

$$\overrightarrow{B_2} = \overrightarrow{T_2} \times \overrightarrow{N_2} \tag{4.77}$$

$$\begin{aligned} \overrightarrow{B_2} = r \cdot & \begin{pmatrix} -a \cdot \sin\alpha + \sin\alpha \cdot \cos(a \cdot \alpha) + a \cdot \cos\alpha \cdot \sin(a \cdot \alpha) \\ a \cdot \cos\alpha - \cos\alpha \cdot \cos(a \cdot \alpha) + a \cdot \sin\alpha \cdot \sin(a \cdot \alpha) \\ -a \cdot \cos(a \cdot \alpha) \end{pmatrix} \\ \times r \cdot & \begin{pmatrix} a \cdot \cos\alpha - a \cdot \cos\alpha + \cos\alpha \cdot \cos(a \cdot \alpha) \\ a \cdot \sin\alpha - a \cdot \sin\alpha + \sin\alpha \cdot \cos(a \cdot \alpha) \\ \sin(a \cdot \alpha) \end{pmatrix} \end{aligned} \tag{4.78}$$

$$\overrightarrow{B_2} = r^2 \cdot \left(\begin{pmatrix} -a \cdot \sin\alpha + \sin\alpha \cdot \cos(a \cdot \alpha) + a \cdot \cos\alpha \cdot \sin(a \cdot \alpha) \\ a \cdot \cos\alpha - \cos\alpha \cdot \cos(a \cdot \alpha) + a \cdot \sin\alpha \cdot \sin(a \cdot \alpha) \\ -a \cdot \cos(a \cdot \alpha) \end{pmatrix} \times \begin{pmatrix} a \cdot \cos\alpha - a \cdot \cos\alpha + \cos\alpha \cdot \cos(a \cdot \alpha) \\ a \cdot \sin\alpha - a \cdot \sin\alpha + \sin\alpha \cdot \cos(a \cdot \alpha) \\ \sin(a \cdot \alpha) \end{pmatrix} \right) \tag{4.79}$$

$$\vec{B_2} = r^2 \cdot \begin{pmatrix} (a \cdot \cos\alpha - \cos\alpha \cdot \cos(a \cdot \alpha) + a \cdot \sin\alpha \cdot \sin(a \cdot \alpha)) \cdot (\sin(a \cdot \alpha)) - \\ -(a \cdot \sin\alpha - a \cdot \sin\alpha + \sin\alpha \cdot \cos(a \cdot \alpha)) \cdot (-a \cdot \cos(a \cdot \alpha)) \\ (-a \cdot \cos(a \cdot \alpha)) \cdot (a \cdot \cos\alpha - a \cdot \cos\alpha + \cos\alpha \cdot \cos(a \cdot \alpha)) - \\ -(\sin(a \cdot \alpha)) \cdot (-a \cdot \sin\alpha + \sin\alpha \cdot \cos(a \cdot \alpha) + a \cdot \cos\alpha \cdot \sin(a \cdot \alpha)) \\ (-a \cdot \sin\alpha + \sin\alpha \cdot \cos(a \cdot \alpha) + a \cdot \cos\alpha \cdot \sin(a \cdot \alpha)) \cdot (a \cdot \sin\alpha - a \cdot \sin\alpha + \sin\alpha \cdot \cos(a \cdot \alpha)) - \\ -(a \cdot \cos\alpha - a \cdot \cos\alpha + \cos\alpha \cdot \cos(a \cdot \alpha)) \cdot (a \cdot \cos\alpha - \cos\alpha \cdot \cos(a \cdot \alpha) + a \cdot \sin\alpha \cdot \sin(a \cdot \alpha)) \end{pmatrix} \tag{4.80}$$

$$\vec{B_2} = r^2 \cdot \begin{pmatrix} (a \cdot \cos\alpha \cdot (\sin(a \cdot \alpha)) - \cos\alpha \cdot \cos(a \cdot \alpha) \cdot (\sin(a \cdot \alpha)) + a \cdot \sin\alpha \cdot (\sin(a \cdot \alpha)) \cdot \sin(a \cdot \alpha)) - \\ -(a \cdot \sin\alpha \cdot (-a \cdot \cos(a \cdot \alpha)) - a \cdot \sin\alpha \cdot (-a \cdot \cos(a \cdot \alpha)) + \sin\alpha \cdot \cos(a \cdot \alpha) \cdot (-a \cdot \cos(a \cdot \alpha))) \\ (-a \cdot \cos(a \cdot \alpha)) \cdot a \cdot \cos\alpha - (-a \cdot \cos(a \cdot \alpha)) \cdot a \cdot \cos\alpha + (-a \cdot \cos(a \cdot \alpha)) \cdot \cos\alpha \cdot \cos(a \cdot \alpha)) - \\ -(-(\sin(a \cdot \alpha)) \cdot a \cdot \sin\alpha + \sin\alpha \cdot (\sin(a \cdot \alpha)) \cdot \cos(a \cdot \alpha) + a \cdot (\sin(a \cdot \alpha)) \cdot \cos\alpha \cdot \sin(a \cdot \alpha)) \\ \begin{pmatrix} -a \cdot \sin\alpha \cdot (a \cdot \sin\alpha - a \cdot \sin\alpha + \sin\alpha \cdot \cos(a \cdot \alpha)) + \\ + \sin\alpha \cdot \cos(a \cdot \alpha) \cdot (a \cdot \sin\alpha - a \cdot \sin\alpha + \sin\alpha \cdot \cos(a \cdot \alpha)) + \\ + a \cdot \cos\alpha \cdot \sin(a \cdot \alpha) \cdot (a \cdot \sin\alpha - a \cdot \sin\alpha + \sin\alpha \cdot \cos(a \cdot \alpha)) \end{pmatrix} - \\ - \begin{pmatrix} a \cdot \cos\alpha \cdot (a \cdot \cos\alpha - \cos\alpha \cdot \cos(a \cdot \alpha) + a \cdot \sin\alpha \cdot \sin(a \cdot \alpha)) - \\ -a \cdot \cos\alpha \cdot (a \cdot \cos\alpha - \cos\alpha \cdot \cos(a \cdot \alpha) + a \cdot \sin\alpha \cdot \sin(a \cdot \alpha)) + \\ + \cos\alpha \cdot \cos(a \cdot \alpha) \cdot (a \cdot \cos\alpha - \cos\alpha \cdot \cos(a \cdot \alpha) + a \cdot \sin\alpha \cdot \sin(a \cdot \alpha)) \end{pmatrix} \end{pmatrix} \tag{4.81}$$

$$\vec{B_2} = r^2 \cdot \begin{pmatrix} -\cos\alpha \cdot \sin(a \cdot \alpha) \cdot \cos(a \cdot \alpha) + a \cdot \sin\alpha + a \cdot \cos\alpha \cdot \sin(a \cdot \alpha) \\ -\sin\alpha \cdot \sin(a \cdot \alpha) \cdot \cos(a \cdot \alpha) - a \cdot \cos\alpha + a \cdot \sin\alpha \cdot \sin(a \cdot \alpha) \\ \cos(a \cdot \alpha) \cdot \cos(a \cdot \alpha) - a \cdot \cos(a \cdot \alpha) \end{pmatrix} \tag{4.82}$$

From the electric or magnetic vector field of the aforementioned form, only a magnetic or electric field of the corresponding form can arise if the original field changes over time.

$$\frac{d}{d\alpha} \begin{pmatrix} x \\ y \\ z \end{pmatrix} = r \cdot \begin{pmatrix} -a \cdot \sin\alpha - \sin\alpha \cdot \cos(a \cdot \alpha) - a \cdot \cos\alpha \cdot \sin(a \cdot \alpha) \\ a \cdot \cos\alpha + \cos\alpha \cdot \cos(a \cdot \alpha) - a \cdot \sin\alpha \cdot \sin(a \cdot \alpha) \\ a \cdot \cos(a \cdot \alpha) \end{pmatrix} \cdot \sin(\omega \cdot t) \tag{4.83}$$

This temporal change, and thus oscillation, occurs because the time sequence of the varying electric field strength leads to a time-varying electric displacement density, which in turn generates a time-varying magnetic field strength. This then transitions into a time-varying magnetic flux density, whose temporal change ultimately produces a time-varying electric field strength again. The time shifts between the individual stages and the 180° phase shift together determine the period of this natural resonance.

The initial configuration described, in which two helices with a pitch of 45° wind around a cylinder—representing the field lines of the electric and magnetic fields and fulfilling the properties described by Maxwell's equations—no longer satisfies these properties when the cylinder is bent into a toroid.

On the outside of the toroid, the field lines become steeper, while on the inside they become flatter; that is, the two opposing vectors representing the electric and magnetic fields, respectively, form an angle $\vartheta \neq 90°$ with each other.

This deviation in angle gives rise to a potential that causes two additional oscillations in the toroid:

- The lateral oscillation (transverse to the carrier circle of the toroid) leads to an alternating electromagnetic field in the external region, which is the cause of the gravitational force.
- The longitudinal oscillation (in the direction of the carrier circle of the toroid) results, on average, in a static electric field that appears externally as either positive or negative and is the cause of electric charge, provided the oscillation amplitude is sufficiently large.

The deviation angle γ can be calculated from a tangent angle and the negative binormal angle of the corresponding other triad, i.e., T_1 and $-B_2$ or T_2 and $-B_1$.

$$\cos\gamma = \frac{\vec{T_1}\cdot\left(-\vec{B_2}\right)}{\left|\vec{T_1}\right|\left|\left(-\vec{B_2}\right)\right|} \tag{4.84}$$

$$\gamma = \arccos\frac{\vec{T_1}\cdot\left(-\vec{B_2}\right)}{\left|\vec{T_1}\right|\left|\left(-\vec{B_2}\right)\right|} \tag{4.85}$$

$$\gamma = \arccos\frac{r\cdot\begin{pmatrix}-a\cdot\sin\alpha-\sin\alpha\cdot\cos(a\cdot\alpha)-a\cdot\cos\alpha\cdot\sin(a\cdot\alpha)\\ a\cdot\cos\alpha+\cos\alpha\cdot\cos(a\cdot\alpha)-a\cdot\sin\alpha\cdot\sin(a\cdot\alpha)\\ a\cdot\cos(a\cdot\alpha)\end{pmatrix}\cdot r^2\cdot(-1)\cdot\begin{pmatrix}-\cos\alpha\cdot\sin(a\cdot\alpha)\cdot\cos(a\cdot\alpha)+a\cdot\sin\alpha+a\cdot\cos\alpha\cdot\sin(a\cdot\alpha)\\ -\sin\alpha\cdot\sin(a\cdot\alpha)\cdot\cos(a\cdot\alpha)-a\cdot\cos\alpha+a\cdot\sin\alpha\cdot\sin(a\cdot\alpha)\\ \cos(a\cdot\alpha)\cdot\cos(a\cdot\alpha)-a\cdot\cos(a\cdot\alpha)\end{pmatrix}}{\left|r\cdot\begin{pmatrix}-a\cdot\sin\alpha-\sin\alpha\cdot\cos(a\cdot\alpha)-a\cdot\cos\alpha\cdot\sin(a\cdot\alpha)\\ a\cdot\cos\alpha+\cos\alpha\cdot\cos(a\cdot\alpha)-a\cdot\sin\alpha\cdot\sin(a\cdot\alpha)\\ a\cdot\cos(a\cdot\alpha)\end{pmatrix}\right|\cdot\left|r^2\cdot(-1)\cdot\begin{pmatrix}-\cos\alpha\cdot\sin(a\cdot\alpha)\cdot\cos(a\cdot\alpha)+a\cdot\sin\alpha+a\cdot\cos\alpha\cdot\sin(a\cdot\alpha)\\ -\sin\alpha\cdot\sin(a\cdot\alpha)\cdot\cos(a\cdot\alpha)-a\cdot\cos\alpha+a\cdot\sin\alpha\cdot\sin(a\cdot\alpha)\\ \cos(a\cdot\alpha)\cdot\cos(a\cdot\alpha)-a\cdot\cos(a\cdot\alpha)\end{pmatrix}\right|} \tag{4.86}$$

$$\gamma = \arccos \frac{\begin{pmatrix} -a \cdot \sin\alpha - \sin\alpha \cdot \cos(a \cdot \alpha) - a \cdot \cos\alpha \cdot \sin(a \cdot \alpha) \\ a \cdot \cos\alpha + \cos\alpha \cdot \cos(a \cdot \alpha) - a \cdot \sin\alpha \cdot \sin(a \cdot \alpha) \\ a \cdot \cos(a \cdot \alpha) \end{pmatrix} \cdot \begin{pmatrix} \cos\alpha \cdot \sin(a \cdot \alpha) \cdot \cos(a \cdot \alpha) - a \cdot \sin\alpha - a \cdot \cos\alpha \cdot \sin(a \cdot \alpha) \\ \sin\alpha \cdot \sin(a \cdot \alpha) \cdot \cos(a \cdot \alpha) + a \cdot \cos\alpha - a \cdot \sin\alpha \cdot \sin(a \cdot \alpha) \\ -\cos(a \cdot \alpha) \cdot \cos(a \cdot \alpha) + a \cdot \cos(a \cdot \alpha) \end{pmatrix}}{\left| \begin{pmatrix} -a \cdot \sin\alpha - \sin\alpha \cdot \cos(a \cdot \alpha) - a \cdot \cos\alpha \cdot \sin(a \cdot \alpha) \\ a \cdot \cos\alpha + \cos\alpha \cdot \cos(a \cdot \alpha) - a \cdot \sin\alpha \cdot \sin(a \cdot \alpha) \\ a \cdot \cos(a \cdot \alpha) \end{pmatrix} \right| \cdot \left| \begin{pmatrix} \cos\alpha \cdot \sin(a \cdot \alpha) \cdot \cos(a \cdot \alpha) - a \cdot \sin\alpha - a \cdot \cos\alpha \cdot \sin(a \cdot \alpha) \\ \sin\alpha \cdot \sin(a \cdot \alpha) \cdot \cos(a \cdot \alpha) + a \cdot \cos\alpha - a \cdot \sin\alpha \cdot \sin(a \cdot \alpha) \\ -\cos(a \cdot \alpha) \cdot \cos(a \cdot \alpha) + a \cdot \cos(a \cdot \alpha) \end{pmatrix} \right|} \tag{4.87}$$

$$\gamma = \arccos \frac{\begin{array}{l} (-a \cdot \sin\alpha - \sin\alpha \cdot \cos(a \cdot \alpha) - a \cdot \cos\alpha \cdot \sin(a \cdot \alpha)) \\ \cdot(\cos\alpha \cdot \sin(a \cdot \alpha) \cdot \cos(a \cdot \alpha) - a \cdot \sin\alpha - a \cdot \cos\alpha \cdot \sin(a \cdot \alpha)) \\ +(a \cdot \cos\alpha + \cos\alpha \cdot \cos(a \cdot \alpha) - a \cdot \sin\alpha \cdot \sin(a \cdot \alpha)) \\ \cdot(\sin\alpha \cdot \sin(a \cdot \alpha) \cdot \cos(a \cdot \alpha) + a \cdot \cos\alpha - a \cdot \sin\alpha \cdot \sin(a \cdot \alpha)) \\ +(a \cdot \cos(a \cdot \alpha)) \cdot (-\cos(a \cdot \alpha) \cdot \cos(a \cdot \alpha) + a \cdot \cos(a \cdot \alpha)) \end{array}}{\sqrt{\begin{array}{l} (-a \cdot \sin\alpha - \sin\alpha \cdot \cos(a \cdot \alpha) - a \cdot \cos\alpha \cdot \sin(a \cdot \alpha)) \\ \cdot(-a \cdot \sin\alpha - \sin\alpha \cdot \cos(a \cdot \alpha) - a \cdot \cos\alpha \cdot \sin(a \cdot \alpha)) \\ +(a \cdot \cos\alpha + \cos\alpha \cdot \cos(a \cdot \alpha) - a \cdot \sin\alpha \cdot \sin(a \cdot \alpha)) \\ \cdot(a \cdot \cos\alpha + \cos\alpha \cdot \cos(a \cdot \alpha) - a \cdot \sin\alpha \cdot \sin(a \cdot \alpha)) \\ +(a \cdot \cos(a \cdot \alpha)) \cdot (a \cdot \cos(a \cdot \alpha)) \end{array}} \cdot \sqrt{\begin{array}{l} (\cos\alpha \cdot \sin(a \cdot \alpha) \cdot \cos(a \cdot \alpha) - a \cdot \sin\alpha - a \cdot \cos\alpha \cdot \sin(a \cdot \alpha)) \\ \cdot(\cos\alpha \cdot \sin(a \cdot \alpha) \cdot \cos(a \cdot \alpha) - a \cdot \sin\alpha - a \cdot \cos\alpha \cdot \sin(a \cdot \alpha)) \\ +(\sin\alpha \cdot \sin(a \cdot \alpha) \cdot \cos(a \cdot \alpha) + a \cdot \cos\alpha - a \cdot \sin\alpha \cdot \sin(a \cdot \alpha)) \\ \cdot(\sin\alpha \cdot \sin(a \cdot \alpha) \cdot \cos(a \cdot \alpha) + a \cdot \cos\alpha - a \cdot \sin\alpha \cdot \sin(a \cdot \alpha)) \\ +(-\cos(a \cdot \alpha) \cdot \cos(a \cdot \alpha) + a \cdot \cos(a \cdot \alpha)) \cdot (-\cos(a \cdot \alpha) \cdot \cos(a \cdot \alpha) + a \cdot \cos(a \cdot \alpha)) \end{array}}} \tag{4.88}$$

$$
\begin{aligned}
&(-a \cdot \sin\alpha - \sin\alpha \cdot \cos(a \cdot \alpha) - a \cdot \cos\alpha \cdot \sin(a \cdot \alpha)) \\
&\cdot (\cos\alpha \cdot \sin(a \cdot \alpha) \cdot \cos(a \cdot \alpha) - a \cdot \sin\alpha - a \cdot \cos\alpha \cdot \sin(a \cdot \alpha)) \\
&+ (a \cdot \cos\alpha + \cos\alpha \cdot \cos(a \cdot \alpha) - a \cdot \sin\alpha \cdot \sin(a \cdot \alpha)) \\
&\cdot (\sin\alpha \cdot \sin(a \cdot \alpha) \cdot \cos(a \cdot \alpha) + a \cdot \cos\alpha - a \cdot \sin\alpha \cdot \sin(a \cdot \alpha)) \\
&+ (a \cdot \cos(a \cdot \alpha)) \cdot (-\cos(a \cdot \alpha) \cdot \cos(a \cdot \alpha) + a \cdot \cos(a \cdot \alpha))
\end{aligned}
\tag{4.89a}
$$

$$
\begin{aligned}
= &\begin{pmatrix}
(-a \cdot \sin\alpha - \sin\alpha \cdot \cos(a \cdot \alpha) - a \cdot \cos\alpha \cdot \sin(a \cdot \alpha)) \cdot \cos\alpha \cdot \sin(a \cdot \alpha) \cdot \cos(a \cdot \alpha) \\
-(-a \cdot \sin\alpha - \sin\alpha \cdot \cos(a \cdot \alpha) - a \cdot \cos\alpha \cdot \sin(a \cdot \alpha)) \cdot a \cdot \sin\alpha \\
-(-a \cdot \sin\alpha - \sin\alpha \cdot \cos(a \cdot \alpha) - a \cdot \cos\alpha \cdot \sin(a \cdot \alpha)) \cdot a \cdot \cos\alpha \cdot \sin(a \cdot \alpha)
\end{pmatrix} \\
+ &\begin{pmatrix}
(a \cdot \cos\alpha + \cos\alpha \cdot \cos(a \cdot \alpha) - a \cdot \sin\alpha \cdot \sin(a \cdot \alpha)) \cdot \sin\alpha \cdot \sin(a \cdot \alpha) \cdot \cos(a \cdot \alpha) \\
+(a \cdot \cos\alpha + \cos\alpha \cdot \cos(a \cdot \alpha) - a \cdot \sin\alpha \cdot \sin(a \cdot \alpha)) \cdot a \cdot \cos\alpha \\
-(a \cdot \cos\alpha + \cos\alpha \cdot \cos(a \cdot \alpha) - a \cdot \sin\alpha \cdot \sin(a \cdot \alpha)) \cdot a \cdot \sin\alpha \cdot \sin(a \cdot \alpha)
\end{pmatrix} \\
+ &(-(a \cdot \cos(a \cdot \alpha)) \cdot \cos(a \cdot \alpha) \cdot \cos(a \cdot \alpha) + (a \cdot \cos(a \cdot \alpha)) \cdot a \cdot \cos(a \cdot \alpha))
\end{aligned}
\tag{4.89b}
$$

$$
\begin{aligned}
= &\begin{pmatrix}
\begin{pmatrix}
-a \cdot \sin\alpha \cdot \cos\alpha \cdot \sin(a \cdot \alpha) \cdot \cos(a \cdot \alpha) \\
-\sin\alpha \cdot \cos(a \cdot \alpha) \cdot \cos\alpha \cdot \sin(a \cdot \alpha) \cdot \cos(a \cdot \alpha) \\
-a \cdot \cos\alpha \cdot \sin(a \cdot \alpha) \cdot \cos\alpha \cdot \sin(a \cdot \alpha) \cdot \cos(a \cdot \alpha)
\end{pmatrix} \\
-(-a \cdot \sin\alpha \cdot a \cdot \sin\alpha - \sin\alpha \cdot \cos(a \cdot \alpha) \cdot a \cdot \sin\alpha - a \cdot \cos\alpha \cdot \sin(a \cdot \alpha) \cdot a \cdot \sin\alpha) \\
-\begin{pmatrix}
-a \cdot \sin\alpha \cdot a \cdot \cos\alpha \cdot \sin(a \cdot \alpha) - \sin\alpha \cdot \cos(a \cdot \alpha) \cdot a \cdot \cos\alpha \cdot \sin(a \cdot \alpha) \\
-a \cdot \cos\alpha \cdot \sin(a \cdot \alpha) \cdot a \cdot \cos\alpha \cdot \sin(a \cdot \alpha)
\end{pmatrix}
\end{pmatrix} \\
+ &\begin{pmatrix}
\begin{pmatrix}
a \cdot \cos\alpha \cdot \sin\alpha \cdot \sin(a \cdot \alpha) \cdot \cos(a \cdot \alpha) \\
+\cos\alpha \cdot \cos(a \cdot \alpha) \cdot \sin\alpha \cdot \sin(a \cdot \alpha) \cdot \cos(a \cdot \alpha) \\
-a \cdot \sin\alpha \cdot \sin(a \cdot \alpha) \cdot \sin\alpha \cdot \sin(a \cdot \alpha) \cdot \cos(a \cdot \alpha)
\end{pmatrix} \\
+(a \cdot \cos\alpha \cdot a \cdot \cos\alpha + \cos\alpha \cdot \cos(a \cdot \alpha) \cdot a \cdot \cos\alpha - a \cdot \sin\alpha \cdot \sin(a \cdot \alpha) \cdot a \cdot \cos\alpha) \\
-\begin{pmatrix}
a \cdot \cos\alpha \cdot a \cdot \sin\alpha \cdot \sin(a \cdot \alpha) + \cos\alpha \cdot \cos(a \cdot \alpha) \cdot a \cdot \sin\alpha \cdot \sin(a \cdot \alpha) \\
-a \cdot \sin\alpha \cdot \sin(a \cdot \alpha) \cdot a \cdot \sin\alpha \cdot \sin(a \cdot \alpha)
\end{pmatrix}
\end{pmatrix} \\
- &a \cdot \cos(a \cdot \alpha) \cdot \cos(a \cdot \alpha) \cdot \cos(a \cdot \alpha) + a \cdot \cos(a \cdot \alpha) \cdot a \cdot \cos(a \cdot \alpha)
\end{aligned}
\tag{4.89c}
$$

$$
\begin{aligned}
&= -a \cdot \sin\alpha \cdot \cos\alpha \cdot \sin(a\cdot\alpha) \cdot \cos(a\cdot\alpha) - \sin\alpha \cdot \cos(a\cdot\alpha) \cdot \cos\alpha \cdot \sin(a\cdot\alpha) \cdot \cos(a\cdot\alpha) \\
&\quad - a \cdot \cos\alpha \cdot \sin(a\cdot\alpha) \cdot \cos\alpha \cdot \sin(a\cdot\alpha) \cdot \cos(a\cdot\alpha) \\
&\quad + a \cdot \sin\alpha \cdot a \cdot \sin\alpha + \sin\alpha \cdot \cos(a\cdot\alpha) \cdot a \cdot \sin\alpha + a \cdot \cos\alpha \cdot \sin(a\cdot\alpha) \cdot a \cdot \sin\alpha \\
&\quad + a \cdot \sin\alpha \cdot a \cdot \cos\alpha \cdot \sin(a\cdot\alpha) + \sin\alpha \cdot \cos(a\cdot\alpha) \cdot a \cdot \cos\alpha \cdot \sin(a\cdot\alpha) \\
&\quad + a \cdot \cos\alpha \cdot \sin(a\cdot\alpha) \cdot a \cdot \cos\alpha \cdot \sin(a\cdot\alpha) \\
&\quad + a \cdot \cos\alpha \cdot \sin\alpha \cdot \sin(a\cdot\alpha) \cdot \cos(a\cdot\alpha) \\
&\quad + \cos\alpha \cdot \cos(a\cdot\alpha) \cdot \sin\alpha \cdot \sin(a\cdot\alpha) \cdot \cos(a\cdot\alpha) \\
&\quad - a \cdot \sin\alpha \cdot \sin(a\cdot\alpha) \cdot \sin\alpha \cdot \sin(a\cdot\alpha) \cdot \cos(a\cdot\alpha) \\
&\quad + a \cdot \cos\alpha \cdot a \cdot \cos\alpha + \cos\alpha \cdot \cos(a\cdot\alpha) \cdot a \cdot \cos\alpha - a \cdot \sin\alpha \cdot \sin(a\cdot\alpha) \cdot a \cdot \cos\alpha \\
&\quad - a \cdot \cos\alpha \cdot a \cdot \sin\alpha \cdot \sin(a\cdot\alpha) - \cos\alpha \cdot \cos(a\cdot\alpha) \cdot a \cdot \sin\alpha \cdot \sin(a\cdot\alpha) \\
&\quad + a \cdot \sin\alpha \cdot \sin(a\cdot\alpha) \cdot a \cdot \sin\alpha \cdot \sin(a\cdot\alpha) \\
&\quad - a \cdot \cos(a\cdot\alpha) \cdot \cos(a\cdot\alpha) \cdot \cos(a\cdot\alpha) + a \cdot \cos(a\cdot\alpha) \cdot a \cdot \cos(a\cdot\alpha)
\end{aligned}
\tag{4.89d}
$$

$$
\begin{aligned}
&= -a \cdot \sin\alpha \cdot \cos\alpha \cdot \sin(a\cdot\alpha) \cdot \cos(a\cdot\alpha) - \sin\alpha \cdot \cos\alpha \cdot \sin(a\cdot\alpha) \cdot \cos(a\cdot\alpha) \cdot \cos(a\cdot\alpha) \\
&\quad - a \cdot \cos\alpha \cdot \cos\alpha \cdot \sin(a\cdot\alpha) \cdot \sin(a\cdot\alpha) \cdot \cos(a\cdot\alpha) \\
&\quad + a \cdot a \cdot \sin\alpha \cdot \sin\alpha + a \cdot \sin\alpha \cdot \sin\alpha \cdot \cos(a\cdot\alpha) + a \cdot a \cdot \sin\alpha \cdot \cos\alpha \cdot \sin(a\cdot\alpha) \\
&\quad + a \cdot a \cdot \sin\alpha \cdot \cos\alpha \cdot \sin(a\cdot\alpha) + a \cdot \sin\alpha \cdot \cos\alpha \cdot \sin(a\cdot\alpha) \cdot \cos(a\cdot\alpha) \\
&\quad + a \cdot a \cdot \cos\alpha \cdot \cos\alpha \cdot \sin(a\cdot\alpha) \cdot \sin(a\cdot\alpha) \\
&\quad + a \cdot \sin\alpha \cdot \cos\alpha \cdot \sin(a\cdot\alpha) \cdot \cos(a\cdot\alpha) \\
&\quad + \sin\alpha \cdot \cos\alpha \cdot \sin(a\cdot\alpha) \cdot \cos(a\cdot\alpha) \cdot \cos(a\cdot\alpha) \\
&\quad - a \cdot \sin\alpha \cdot \sin\alpha \cdot \sin(a\cdot\alpha) \cdot \sin(a\cdot\alpha) \cdot \cos(a\cdot\alpha) \\
&\quad + a \cdot a \cdot \cos\alpha \cdot \cos\alpha + a \cdot \cos\alpha \cdot \cos\alpha \cdot \cos(a\cdot\alpha) - a \cdot a \cdot \sin\alpha \cdot \cos\alpha \cdot \sin(a\cdot\alpha) \\
&\quad - a \cdot a \cdot \sin\alpha \cdot \cos\alpha \cdot \sin(a\cdot\alpha) - a \cdot \sin\alpha \cdot \cos\alpha \cdot \sin(a\cdot\alpha) \cdot \cos(a\cdot\alpha) \\
&\quad + a \cdot a \cdot \sin\alpha \cdot \sin\alpha \cdot \sin(a\cdot\alpha) \cdot \sin(a\cdot\alpha) \\
&\quad - a \cdot \cos(a\cdot\alpha) \cdot \cos(a\cdot\alpha) \cdot \cos(a\cdot\alpha) + a \cdot a \cdot \cos(a\cdot\alpha) \cdot \cos(a\cdot\alpha)
\end{aligned}
\tag{4.89e}
$$

$$
= 2 \cdot a^2 (\text{Numerator}) \tag{4.89f}
$$

$$
\sqrt{
\begin{aligned}
&(-a \cdot \sin\alpha - \sin\alpha \cdot \cos(a\cdot\alpha) - a \cdot \cos\alpha \cdot \sin(a\cdot\alpha)) \\
&\cdot(-a \cdot \sin\alpha - \sin\alpha \cdot \cos(a\cdot\alpha) - a \cdot \cos\alpha \cdot \sin(a\cdot\alpha)) \\
&+(a \cdot \cos\alpha + \cos\alpha \cdot \cos(a\cdot\alpha) - a \cdot \sin\alpha \cdot \sin(a\cdot\alpha)) \\
&\cdot(a \cdot \cos\alpha + \cos\alpha \cdot \cos(a\cdot\alpha) - a \cdot \sin\alpha \cdot \sin(a\cdot\alpha)) \\
&+(a \cdot \cos(a\cdot\alpha)) \cdot (a \cdot \cos(a\cdot\alpha))
\end{aligned}
}
\tag{4.90a}
$$

$$= \sqrt{\begin{array}{l} \left(\begin{array}{l} -(-a\cdot\sin\alpha\cdot a\cdot\sin\alpha - a\cdot\sin\alpha\cdot\sin\alpha\cdot\cos(a\cdot\alpha) - a\cdot\sin\alpha\cdot a\cdot\cos\alpha\cdot\sin(a\cdot\alpha)) \\ -\left(\begin{array}{l} -\sin\alpha\cdot\cos(a\cdot\alpha)\cdot a\cdot\sin\alpha - \sin\alpha\cdot\cos(a\cdot\alpha)\cdot\sin\alpha\cdot\cos(a\cdot\alpha) \\ -\sin\alpha\cdot\cos(a\cdot\alpha)\cdot a\cdot\cos\alpha\cdot\sin(a\cdot\alpha) \end{array}\right) \\ -\left(\begin{array}{l} -a\cdot\cos\alpha\cdot\sin(a\cdot\alpha)\cdot a\cdot\sin\alpha - a\cdot\cos\alpha\cdot\sin(a\cdot\alpha)\cdot\sin\alpha\cdot\cos(a\cdot\alpha) \\ -a\cdot\cos\alpha\cdot\sin(a\cdot\alpha)\cdot a\cdot\cos\alpha\cdot\sin(a\cdot\alpha) \end{array}\right) \end{array}\right) \\ +\left(\begin{array}{l} (a\cdot\cos\alpha\cdot a\cdot\cos\alpha + a\cdot\cos\alpha\cdot\cos\alpha\cdot\cos(a\cdot\alpha) - a\cdot\cos\alpha\cdot a\cdot\sin\alpha\cdot\sin(a\cdot\alpha)) \\ +\left(\begin{array}{l} \cos\alpha\cdot\cos(a\cdot\alpha)\cdot a\cdot\cos\alpha + \cos\alpha\cdot\cos(a\cdot\alpha)\cdot\cos\alpha\cdot\cos(a\cdot\alpha) \\ -\cos\alpha\cdot\cos(a\cdot\alpha)\cdot a\cdot\sin\alpha\cdot\sin(a\cdot\alpha) \end{array}\right) \\ -\left(\begin{array}{l} a\cdot\sin\alpha\cdot\sin(a\cdot\alpha)\cdot a\cdot\cos\alpha + a\cdot\sin\alpha\cdot\sin(a\cdot\alpha)\cdot\cos\alpha\cdot\cos(a\cdot\alpha) \\ -a\cdot\sin\alpha\cdot\sin(a\cdot\alpha)\cdot a\cdot\sin\alpha\cdot\sin(a\cdot\alpha) \end{array}\right) \end{array}\right) \\ +(a\cdot\cos(a\cdot\alpha))\cdot(a\cdot\cos(a\cdot\alpha)) \end{array}} \tag{4.90b}$$

$$= \sqrt{\begin{array}{l} a\cdot\sin\alpha\cdot a\cdot\sin\alpha + a\cdot\sin\alpha\cdot\sin\alpha\cdot\cos(a\cdot\alpha) + a\cdot\sin\alpha\cdot a\cdot\cos\alpha\cdot\sin(a\cdot\alpha) \\ +\sin\alpha\cdot\cos(a\cdot\alpha)\cdot a\cdot\sin\alpha + \sin\alpha\cdot\cos(a\cdot\alpha)\cdot\sin\alpha\cdot\cos(a\cdot\alpha) \\ +\sin\alpha\cdot\cos(a\cdot\alpha)\cdot a\cdot\cos\alpha\cdot\sin(a\cdot\alpha) \\ +a\cdot\cos\alpha\cdot\sin(a\cdot\alpha)\cdot a\cdot\sin\alpha + a\cdot\cos\alpha\cdot\sin(a\cdot\alpha)\cdot\sin\alpha\cdot\cos(a\cdot\alpha) \\ +a\cdot\cos\alpha\cdot\sin(a\cdot\alpha)\cdot a\cdot\cos\alpha\cdot\sin(a\cdot\alpha) \\ +a\cdot\cos\alpha\cdot a\cdot\cos\alpha + a\cdot\cos\alpha\cdot\cos\alpha\cdot\cos(a\cdot\alpha) - a\cdot\cos\alpha\cdot a\cdot\sin\alpha\cdot\sin(a\cdot\alpha) \\ +\cos\alpha\cdot\cos(a\cdot\alpha)\cdot a\cdot\cos\alpha + \cos\alpha\cdot\cos(a\cdot\alpha)\cdot\cos\alpha\cdot\cos(a\cdot\alpha) \\ -\cos\alpha\cdot\cos(a\cdot\alpha)\cdot a\cdot\sin\alpha\cdot\sin(a\cdot\alpha) \\ -a\cdot\sin\alpha\cdot\sin(a\cdot\alpha)\cdot a\cdot\cos\alpha - a\cdot\sin\alpha\cdot\sin(a\cdot\alpha)\cdot\cos\alpha\cdot\cos(a\cdot\alpha) \\ +a\cdot\sin\alpha\cdot\sin(a\cdot\alpha)\cdot a\cdot\sin\alpha\cdot\sin(a\cdot\alpha) \\ +a\cdot\cos(a\cdot\alpha)\cdot a\cdot\cos(a\cdot\alpha) \end{array}} \tag{4.90c}$$

$$= \sqrt{\begin{array}{l} +\sin\alpha\cdot\sin\alpha\cdot\cos(a\cdot\alpha)\cdot\cos(a\cdot\alpha)+\cos\alpha\cdot\cos\alpha\cdot\cos(a\cdot\alpha)\cdot\cos(a\cdot\alpha) \\ +a\cdot\sin\alpha\cdot\sin\alpha\cdot\cos(a\cdot\alpha)+a\cdot\sin\alpha\cdot\sin\alpha\cdot\cos(a\cdot\alpha) \\ +a\cdot\cos\alpha\cdot\cos\alpha\cdot\cos(a\cdot\alpha)+a\cdot\cos\alpha\cdot\cos\alpha\cdot\cos(a\cdot\alpha) \\ +a\cdot\sin\alpha\cdot\cos\alpha\cdot\sin(a\cdot\alpha)\cdot\cos(a\cdot\alpha)-a\cdot\sin\alpha\cdot\cos\alpha\cdot\sin(a\cdot\alpha)\cdot\cos(a\cdot\alpha) \\ +a\cdot\sin\alpha\cdot\cos\alpha\cdot\sin(a\cdot\alpha)\cdot\cos(a\cdot\alpha)-a\cdot\sin\alpha\cdot\cos\alpha\cdot\sin(a\cdot\alpha)\cdot\cos(a\cdot\alpha) \\ +a\cdot a\cdot\sin\alpha\cdot\sin\alpha+a\cdot a\cdot\cos\alpha\cdot\cos\alpha \\ +a\cdot a\cdot\sin\alpha\cdot\cos\alpha\cdot\sin(a\cdot\alpha)+a\cdot a\cdot\sin\alpha\cdot\cos\alpha\cdot\sin(a\cdot\alpha) \\ -a\cdot a\cdot\sin\alpha\cdot\cos\alpha\cdot\sin(a\cdot\alpha)-a\cdot a\cdot\sin\alpha\cdot\cos\alpha\cdot\sin(a\cdot\alpha) \\ +a\cdot a\cdot\sin\alpha\cdot\sin\alpha\cdot\sin(a\cdot\alpha)\cdot\sin(a\cdot\alpha) \\ +a\cdot a\cdot\cos\alpha\cdot\cos\alpha\cdot\sin(a\cdot\alpha)\cdot\sin(a\cdot\alpha) \\ +a\cdot a\cdot\cos(a\cdot\alpha)\cdot\cos(a\cdot\alpha) \end{array}} \tag{4.90d}$$

$$= \sqrt{\begin{array}{l} \cos(a\cdot\alpha)\cdot\cos(a\cdot\alpha) \\ +a\cdot\cos(a\cdot\alpha)+a\cdot\cos(a\cdot\alpha) \\ +a\cdot a+a\cdot a \end{array}} \tag{4.90e}$$

$$= \sqrt{\cos^2(a\cdot\alpha)+2\cdot a\cdot\cos(a\cdot\alpha)+2\cdot a^2} \tag{4.90f}$$

$$= \sqrt{(\cos(a\cdot\alpha)+a)^2+a^2} \tag{4.90g}$$

$$\sqrt{\begin{array}{l} (\cos\alpha\cdot\sin(a\cdot\alpha)\cdot\cos(a\cdot\alpha)-a\cdot\sin\alpha-a\cdot\cos\alpha\cdot\sin(a\cdot\alpha)) \\ \cdot(\cos\alpha\cdot\sin(a\cdot\alpha)\cdot\cos(a\cdot\alpha)-a\cdot\sin\alpha-a\cdot\cos\alpha\cdot\sin(a\cdot\alpha)) \\ +(\sin\alpha\cdot\sin(a\cdot\alpha)\cdot\cos(a\cdot\alpha)+a\cdot\cos\alpha-a\cdot\sin\alpha\cdot\sin(a\cdot\alpha)) \\ \cdot(\sin\alpha\cdot\sin(a\cdot\alpha)\cdot\cos(a\cdot\alpha)+a\cdot\cos\alpha-a\cdot\sin\alpha\cdot\sin(a\cdot\alpha)) \\ +(-\cos(a\cdot\alpha)\cdot\cos(a\cdot\alpha)+a\cdot\cos(a\cdot\alpha))\cdot(-\cos(a\cdot\alpha)\cdot\cos(a\cdot\alpha)+a\cdot\cos(a\cdot\alpha)) \end{array}} \tag{4.91a}$$

$$
= \sqrt{\begin{array}{l}
\begin{pmatrix}
\cos\alpha\cdot\sin(a\cdot\alpha)\cdot\cos(a\cdot\alpha)\cdot(\cos\alpha\cdot\sin(a\cdot\alpha)\cdot\cos(a\cdot\alpha)-a\cdot\sin\alpha-a\cdot\cos\alpha\cdot\sin(a\cdot\alpha)) \\
-a\cdot\sin\alpha\cdot(\cos\alpha\cdot\sin(a\cdot\alpha)\cdot\cos(a\cdot\alpha)-a\cdot\sin\alpha-a\cdot\cos\alpha\cdot\sin(a\cdot\alpha)) \\
-a\cdot\cos\alpha\cdot\sin(a\cdot\alpha)\cdot(\cos\alpha\cdot\sin(a\cdot\alpha)\cdot\cos(a\cdot\alpha)-a\cdot\sin\alpha-a\cdot\cos\alpha\cdot\sin(a\cdot\alpha))
\end{pmatrix} \\
+\begin{pmatrix}
\sin\alpha\cdot\sin(a\cdot\alpha)\cdot\cos(a\cdot\alpha)\cdot(\sin\alpha\cdot\sin(a\cdot\alpha)\cdot\cos(a\cdot\alpha)+a\cdot\cos\alpha-a\cdot\sin\alpha\cdot\sin(a\cdot\alpha)) \\
+a\cdot\cos\alpha\cdot(\sin\alpha\cdot\sin(a\cdot\alpha)\cdot\cos(a\cdot\alpha)+a\cdot\cos\alpha-a\cdot\sin\alpha\cdot\sin(a\cdot\alpha)) \\
-a\cdot\sin\alpha\cdot\sin(a\cdot\alpha)\cdot(\sin\alpha\cdot\sin(a\cdot\alpha)\cdot\cos(a\cdot\alpha)+a\cdot\cos\alpha-a\cdot\sin\alpha\cdot\sin(a\cdot\alpha))
\end{pmatrix} \\
+\begin{pmatrix}
-\cos(a\cdot\alpha)\cdot\cos(a\cdot\alpha)\cdot(-\cos(a\cdot\alpha)\cdot\cos(a\cdot\alpha)+a\cdot\cos(a\cdot\alpha)) \\
+a\cdot\cos(a\cdot\alpha)\cdot(-\cos(a\cdot\alpha)\cdot\cos(a\cdot\alpha)+a\cdot\cos(a\cdot\alpha))
\end{pmatrix}
\end{array}} \tag{4.91b}
$$

$$
= \sqrt{\begin{array}{l}
\begin{pmatrix}
\begin{pmatrix}
\cos\alpha\cdot\sin(a\cdot\alpha)\cdot\cos(a\cdot\alpha)\cdot\cos\alpha\cdot\sin(a\cdot\alpha)\cdot\cos(a\cdot\alpha) \\
-\cos\alpha\cdot\sin(a\cdot\alpha)\cdot\cos(a\cdot\alpha)\cdot a\cdot\sin\alpha-\cos\alpha\cdot\sin(a\cdot\alpha)\cdot\cos(a\cdot\alpha)\cdot a\cdot\cos\alpha\cdot\sin(a\cdot\alpha)
\end{pmatrix} \\
-(a\cdot\sin\alpha\cdot\cos\alpha\cdot\sin(a\cdot\alpha)\cdot\cos(a\cdot\alpha)-a\cdot\sin\alpha\cdot a\cdot\sin\alpha-a\cdot\sin\alpha\cdot a\cdot\cos\alpha\cdot\sin(a\cdot\alpha)) \\
-\begin{pmatrix}
a\cdot\cos\alpha\cdot\sin(a\cdot\alpha)\cdot\cos\alpha\cdot\sin(a\cdot\alpha)\cdot\cos(a\cdot\alpha) \\
-a\cdot\cos\alpha\cdot\sin(a\cdot\alpha)\cdot a\cdot\sin\alpha-a\cdot\cos\alpha\cdot\sin(a\cdot\alpha)\cdot a\cdot\cos\alpha\cdot\sin(a\cdot\alpha)
\end{pmatrix}
\end{pmatrix} \\
+\begin{pmatrix}
\begin{pmatrix}
\sin\alpha\cdot\sin(a\cdot\alpha)\cdot\cos(a\cdot\alpha)\cdot\sin\alpha\cdot\sin(a\cdot\alpha)\cdot\cos(a\cdot\alpha) \\
+\sin\alpha\cdot\sin(a\cdot\alpha)\cdot\cos(a\cdot\alpha)\cdot a\cdot\cos\alpha-\sin\alpha\cdot\sin(a\cdot\alpha)\cdot\cos(a\cdot\alpha)\cdot a\cdot\sin\alpha\cdot\sin(a\cdot\alpha)
\end{pmatrix} \\
+(a\cdot\cos\alpha\cdot\sin\alpha\cdot\sin(a\cdot\alpha)\cdot\cos(a\cdot\alpha)+a\cdot\cos\alpha\cdot a\cdot\cos\alpha-a\cdot\cos\alpha\cdot a\cdot\sin\alpha\cdot\sin(a\cdot\alpha)) \\
-\begin{pmatrix}
a\cdot\sin\alpha\cdot\sin(a\cdot\alpha)\cdot\sin\alpha\cdot\sin(a\cdot\alpha)\cdot\cos(a\cdot\alpha)+a\cdot\sin\alpha\cdot\sin(a\cdot\alpha)\cdot a\cdot\cos\alpha \\
-a\cdot\sin\alpha\cdot\sin(a\cdot\alpha)\cdot a\cdot\sin\alpha\cdot\sin(a\cdot\alpha)
\end{pmatrix}
\end{pmatrix} \\
+\begin{pmatrix}
-(-\cos(a\cdot\alpha)\cdot\cos(a\cdot\alpha)\cdot\cos(a\cdot\alpha)\cdot\cos(a\cdot\alpha)+\cos(a\cdot\alpha)\cdot\cos(a\cdot\alpha)\cdot a\cdot\cos(a\cdot\alpha)) \\
+(-a\cdot\cos(a\cdot\alpha)\cdot\cos(a\cdot\alpha)\cdot\cos(a\cdot\alpha)+a\cdot\cos(a\cdot\alpha)\cdot a\cdot\cos(a\cdot\alpha))
\end{pmatrix}
\end{array}} \tag{4.91c}
$$

$$
= \sqrt{\begin{array}{l}
\cos\alpha\cdot\cos\alpha\cdot\sin(a\cdot\alpha)\cdot\sin(a\cdot\alpha)\cdot\cos(a\cdot\alpha)\cdot\cos(a\cdot\alpha) \\
-a\cdot\sin\alpha\cdot\cos\alpha\cdot\sin(a\cdot\alpha)\cdot\cos(a\cdot\alpha)-a\cdot\cos\alpha\cdot\cos\alpha\cdot\sin(a\cdot\alpha)\cdot\sin(a\cdot\alpha)\cdot\cos(a\cdot\alpha) \\
-a\cdot\sin\alpha\cdot\cos\alpha\cdot\sin(a\cdot\alpha)\cdot\cos(a\cdot\alpha)+a\cdot a\cdot\sin\alpha\cdot\sin\alpha+a\cdot a\cdot\sin\alpha\cdot\cos\alpha\cdot\sin(a\cdot\alpha) \\
-a\cdot\cos\alpha\cdot\cos\alpha\cdot\sin(a\cdot\alpha)\cdot\sin(a\cdot\alpha)\cdot\cos(a\cdot\alpha) \\
+a\cdot a\cdot\sin\alpha\cdot\cos\alpha\cdot\sin(a\cdot\alpha)+a\cdot a\cdot\cos\alpha\cdot\cos\alpha\cdot\sin(a\cdot\alpha)\cdot\sin(a\cdot\alpha) \\
+\sin\alpha\cdot\sin\alpha\cdot\sin(a\cdot\alpha)\cdot\sin(a\cdot\alpha)\cdot\cos(a\cdot\alpha)\cdot\cos(a\cdot\alpha) \\
+a\cdot\sin\alpha\cdot\cos\alpha\cdot\sin(a\cdot\alpha)\cdot\cos(a\cdot\alpha)-a\cdot\sin\alpha\cdot\sin\alpha\cdot\sin(a\cdot\alpha)\cdot\sin(a\cdot\alpha)\cdot\cos(a\cdot\alpha) \\
+a\cdot\sin\alpha\cdot\cos\alpha\cdot\sin(a\cdot\alpha)\cdot\cos(a\cdot\alpha)+a\cdot a\cdot\cos\alpha\cdot\cos\alpha-a\cdot a\cdot\sin\alpha\cdot\cos\alpha\cdot\sin(a\cdot\alpha) \\
-a\cdot\sin\alpha\cdot\sin\alpha\cdot\sin(a\cdot\alpha)\cdot\sin(a\cdot\alpha)\cdot\cos(a\cdot\alpha)-a\cdot a\cdot\sin\alpha\cdot\cos\alpha\cdot\sin(a\cdot\alpha) \\
+a\cdot a\cdot\sin\alpha\cdot\sin\alpha\cdot\sin(a\cdot\alpha)\cdot\sin(a\cdot\alpha) \\
+\cos(a\cdot\alpha)\cdot\cos(a\cdot\alpha)\cdot\cos(a\cdot\alpha)\cdot\cos(a\cdot\alpha)-a\cdot\cos(a\cdot\alpha)\cdot\cos(a\cdot\alpha)\cdot\cos(a\cdot\alpha) \\
-a\cdot\cos(a\cdot\alpha)\cdot\cos(a\cdot\alpha)\cdot\cos(a\cdot\alpha)+a\cdot a\cdot\cos(a\cdot\alpha)\cdot\cos(a\cdot\alpha)
\end{array}} \tag{4.91d}
$$

$$= \sqrt{\begin{array}{l} \cos\alpha \cdot \cos\alpha \cdot \sin(a\cdot\alpha) \cdot \sin(a\cdot\alpha) \cdot \cos(a\cdot\alpha) \cdot \cos(a\cdot\alpha) \\ +\cos(a\cdot\alpha) \cdot \cos(a\cdot\alpha) \cdot \cos(a\cdot\alpha) \cdot \cos(a\cdot\alpha) \\ +\sin\alpha \cdot \sin\alpha \cdot \sin(a\cdot\alpha) \cdot \sin(a\cdot\alpha) \cdot \cos(a\cdot\alpha) \cdot \cos(a\cdot\alpha) \\ -a \cdot \sin\alpha \cdot \cos\alpha \cdot \sin(a\cdot\alpha) \cdot \cos(a\cdot\alpha) - a \cdot \cos\alpha \cdot \cos\alpha \cdot \sin(a\cdot\alpha) \cdot \sin(a\cdot\alpha) \cdot \cos(a\cdot\alpha) \\ -a \cdot \sin\alpha \cdot \cos\alpha \cdot \sin(a\cdot\alpha) \cdot \cos(a\cdot\alpha) - a \cdot \cos(a\cdot\alpha) \cdot \cos(a\cdot\alpha) \cdot \cos(a\cdot\alpha) \\ -a \cdot \cos\alpha \cdot \cos\alpha \cdot \sin(a\cdot\alpha) \cdot \sin(a\cdot\alpha) \cdot \cos(a\cdot\alpha) \\ +a \cdot \sin\alpha \cdot \cos\alpha \cdot \sin(a\cdot\alpha) \cdot \cos(a\cdot\alpha) - a \cdot \sin\alpha \cdot \sin\alpha \cdot \sin(a\cdot\alpha) \cdot \sin(a\cdot\alpha) \cdot \cos(a\cdot\alpha) \\ +a \cdot \sin\alpha \cdot \cos\alpha \cdot \sin(a\cdot\alpha) \cdot \cos(a\cdot\alpha) \\ -a \cdot \cos(a\cdot\alpha) \cdot \cos(a\cdot\alpha) \cdot \cos(a\cdot\alpha) \\ -a \cdot \cos(a\cdot\alpha) \cdot \cos(a\cdot\alpha) \cdot \cos(a\cdot\alpha) \\ +a \cdot a \cdot \sin\alpha \cdot \sin\alpha \cdot \sin(a\cdot\alpha) \cdot \sin(a\cdot\alpha) + a \cdot a \cdot \sin\alpha \cdot \sin\alpha + a \cdot a \cdot \sin\alpha \cdot \cos\alpha \cdot \sin(a\cdot\alpha) \\ +a \cdot a \cdot \sin\alpha \cdot \cos\alpha \cdot \sin(a\cdot\alpha) + a \cdot a \cdot \cos\alpha \cdot \cos\alpha \cdot \sin(a\cdot\alpha) \cdot \sin(a\cdot\alpha) \\ +a \cdot a \cdot \cos\alpha \cdot \cos\alpha - a \cdot a \cdot \sin\alpha \cdot \cos\alpha \cdot \sin(a\cdot\alpha) - a \cdot a \cdot \sin\alpha \cdot \cos\alpha \cdot \sin(a\cdot\alpha) \\ +a \cdot a \cdot \cos(a\cdot\alpha) \cdot \cos(a\cdot\alpha) \end{array}} \tag{4.91e}$$

$$= \sqrt{\begin{array}{l} \cos(a\cdot\alpha) \cdot \cos(a\cdot\alpha) \\ -a \cdot \cos(a\cdot\alpha) - a \cdot \cos(a\cdot\alpha) \\ +a \cdot a + a \cdot a \end{array}} \tag{4.91f}$$

$$= \sqrt{(\cos(a\cdot\alpha) - a)^2 + a^2} \tag{4.91g}$$

Denominator:

$$\sqrt{(\cos(a\cdot\alpha) + a)^2 + a^2} \cdot \sqrt{(\cos(a\cdot\alpha) - a)^2 + a^2} \tag{4.92a}$$

$$= \sqrt{\left((\cos(a\cdot\alpha) + a)^2 + a^2\right) \cdot \left((\cos(a\cdot\alpha) - a)^2 + a^2\right)} \tag{4.92b}$$

$$= \sqrt{\begin{array}{l} \left((\cos(a\cdot\alpha) + a)^2 \cdot (\cos(a\cdot\alpha) - a)^2 + (\cos(a\cdot\alpha) + a)^2 \cdot a^2\right) \\ +\left(a^2 \cdot (\cos(a\cdot\alpha) - a)^2 + a^2 \cdot a^2\right) \end{array}} \tag{4.92c}$$

$$= \sqrt{\left(\cos^2(a\cdot\alpha) - a^2\right)^2 + 2 \cdot a^2 \cdot \left(\cos^2(a\cdot\alpha) + a^2\right) + a^2 \cdot a^2} \tag{4.92d}$$

$$= \sqrt{\cos^4(a \cdot \alpha) + 4 \cdot a^4} \tag{4.92e}$$

$$\gamma = \arccos \frac{2 \cdot a^2}{\sqrt{\cos^4(a \cdot \alpha) + 4 \cdot a^4}} \tag{4.93}$$

The absolute value of the argument of the arccosine must always be less than or equal to 1.

Because:

$$2 \cdot a^2 = \sqrt{4 \cdot a^4} \tag{4.94}$$

and because

$$\cos^4(a \cdot \alpha) > 0 \tag{4.95}$$

and therefore

$$2 \cdot a^2 < \sqrt{\cos^4(a \cdot \alpha) + 4 \cdot a^4}, \tag{4.96}$$

it follows that:

$$\frac{2 \cdot a^2}{\sqrt{\cos^4(a \cdot \alpha) + 4 \cdot a^4}} < 1 \tag{4.97}$$

This argument allows the angle γ to be calculated as a function of a and the angle α.

$$\gamma = \arccos \frac{2 \cdot a^2}{\sqrt{\cos^4(a \cdot \alpha) + 4 \cdot a^4}} \tag{4.98}$$

This deviation angle γ can be used to describe a force that attempts to restore the normal state between the two fields. This force represents potential energy. Through a compensation process, the potential energy is converted into kinetic energy, which leads to a deflection of the same magnitude and thus back to potential energy, resulting in the formation of an oscillator. An estimate of the oscillator's frequency is provided elsewhere.

4.1 The Maxwell Equations

The Maxwell equations still form the fundamental framework for electrical engineering today. With appropriate interpretation, they can be used to describe all behaviors, properties, and phenomena in electrical engineering [3].

4.1.1 Origin of the Maxwell Equations

James Clerk Maxwell developed the system of equations named after him from Ampère's circuital law, Gauss's law, and the law of induction, and supplemented them with the displacement current [4].

4.1.2 The System of Equations in Its Integral Form

$$\oiint_{dV} \vec{D} \cdot d\vec{A} = \iiint_{V} \rho \cdot dV = Q(V) \tag{4.99}$$

The electric displacement density emerging from or entering the surface of a volume is equal to the amount of electric charge within that volume. In other words: Electric fields can have a source.

$$\oiint_{dV} \vec{B} \cdot d\vec{A} = 0 \tag{4.100}$$

The magnetic flux density entering a volume through its surface also exits the surface of the volume. In other words: Magnetic fields are source-free.

$$\oint_{dA} \vec{E} \cdot d\vec{s} = -\iint_{A} \frac{d\vec{B}}{dt} \cdot d\vec{A} \tag{4.101}$$

The integral of the electric field strength E over the line segments ds of the path surrounding the area A, through which a time-varying magnetic flux density B passes, is equal to the total magnetic flux through this area. In other words: If a time-varying magnetic flux is present in a loop, an electric voltage is induced in this loop. The greater the change in magnetic flux per unit time, the greater the induced voltage per unit time.

$$\oint_{dA} \vec{H} \cdot d\vec{s} = \iint_{A} \vec{j_l} \cdot d\vec{A} + \iint_{A} \frac{d\vec{D}}{dt} \cdot d\vec{A} \tag{4.102}$$

The integral of the magnetic field strength H over the line segments ds of the path surrounding the area A, through which a time-varying electric displacement density D or a direct electric current flows, is equal to the total magnetic flux through this area. In other words: If a time-varying or constant electric displacement density or current is present in a loop, a magnetic field strength is generated in this loop.

4.1.3 The System of Equations in Its Differential Form

$$div\,\vec{D} = \vec{\nabla} \cdot \vec{D} = \rho \tag{4.103}$$

The electric displacement field has a source.

$$div\,\vec{B} = \vec{\nabla} \cdot \vec{B} = 0 \tag{4.104}$$

The magnetic flux density field has no source.

$$rot\,\vec{E} = \vec{\nabla} \times \vec{E} = -\frac{d\,\vec{B}}{dt} \tag{4.105}$$

The spatial variations (curls) of the electric field strength are equal to the temporal variation of the magnetic flux density.

$$rot\,\vec{H} = \vec{\nabla} \times \vec{H} = \vec{j_l} + \frac{d\,\vec{D}}{dt} \tag{4.106}$$

The spatial variations (curls) of the magnetic field strength are equal to the sum of the electric current and the temporal variation of the displacement density.

4.1.4 The Constitutive Equations

$$\vec{D} = \varepsilon_0 \cdot \vec{E} + \vec{P} \tag{4.107}$$

The electric displacement density follows the electric field strength over time, increased by any existing potential, where the constant ε_0 applies only to vacuum. In material media, the constant is multiplied by a factor ε_r. A further derivation of this constant can be found in Chap. 6 Mass.

$$\vec{B} = \mu_0 \cdot \vec{H} + \vec{M} \tag{4.108}$$

The magnetic flux density follows the magnetic field strength over time, increased by a magnetic potential, where the constant μ_0 applies only to vacuum. In material media, the constant is multiplied by a factor μ_r. A further derivation of this constant can be found in Chap. 6 Mass.

$$\vec{J} = \sigma \cdot \vec{E} \tag{4.109}$$

The electric current is determined by the electrical conductivity of a volume and the electric field strength within that volume.

4.1.5 The Curl Operator

The curl operator is generally defined as the cross product of the nabla operator with the vector to which the curl is to be applied [5]:

$$rot\,\vec{v} = rot\begin{pmatrix} v_x \\ v_y \\ v_z \end{pmatrix} = \begin{pmatrix} \frac{d}{dx} \\ \frac{d}{dy} \\ \frac{d}{dz} \end{pmatrix} \times \begin{pmatrix} v_x \\ v_y \\ v_z \end{pmatrix} = \begin{pmatrix} \frac{dv_z}{dy} - \frac{dv_y}{dz} \\ \frac{dv_x}{dz} - \frac{dv_z}{dx} \\ \frac{dv_y}{dx} - \frac{dv_x}{dy} \end{pmatrix} \tag{4.110}$$

Applying the curl operator in the cylindrical coordinate system yields the following representation [5]:

$$rot\,\vec{u} = \begin{pmatrix} \frac{1}{r} \cdot \frac{du_z}{d\phi} - \frac{du_\phi}{dz} \\ \frac{du_r}{dz} - \frac{du_z}{dr} \\ \frac{1}{r} \cdot \frac{d(r \cdot u_\phi)}{dr} - \frac{1}{r} \cdot \frac{du_r}{d\phi} \end{pmatrix} \cdot \begin{pmatrix} \vec{e_r} \\ \vec{e_\phi} \\ \vec{e_z} \end{pmatrix} \tag{4.111}$$

The transformation functions between the Cartesian coordinate system and the cylindrical coordinate system for position vectors are [5]:

$$\vec{v} = \vec{v}(r, \varphi, z) = \begin{pmatrix} v_x \\ v_y \\ v_z \end{pmatrix} = \begin{pmatrix} r \cdot \cos\varphi \\ r \cdot \sin\varphi \\ z \end{pmatrix} \tag{4.112}$$

$$\vec{u} = \vec{u}(x, y, z) = \begin{pmatrix} u_r \\ u_\phi \\ u_z \end{pmatrix} = \begin{pmatrix} \sqrt{x^2 + y^2} \\ \arctan \frac{y}{x} \\ z \end{pmatrix} \tag{4.113}$$

and for vectors [Bronstein]:

$$\vec{v}(r, \phi, z) = \begin{pmatrix} v_x \cdot \cos\phi + v_y \cdot \sin\phi \\ -v_x \cdot \sin\phi + v_y \cdot \cos\phi \\ v_z \end{pmatrix} \tag{4.114}$$

$$\vec{v}(x, y, z) = \begin{pmatrix} v_x \\ v_y \\ v_z \end{pmatrix} = \begin{pmatrix} v_r \cdot \cos\phi - v_\phi \cdot \sin\phi \\ v_r \cdot \sin\phi + v_\phi \cdot \cos\phi \\ v_z \end{pmatrix} \tag{4.115}$$

According to [5], the curl operator is

$$\vec{v} = \omega \cdot (\vec{n} \times \vec{r}) \tag{4.116}$$

With

$$\vec{n} = \begin{pmatrix} n_x \\ n_y \\ n_z \end{pmatrix} \tag{4.117}$$

and

$$\vec{r} = \begin{pmatrix} r \cdot \cos\phi \\ r \cdot \sin\phi \\ z \end{pmatrix} \tag{4.118}$$

one obtains

$$\vec{v} = \omega \cdot \left(\begin{pmatrix} n_x \\ n_y \\ n_z \end{pmatrix} \times \begin{pmatrix} r \cdot \cos\phi \\ r \cdot \sin\phi \\ z \end{pmatrix} \right) \tag{4.119}$$

$$\vec{v} = \omega \cdot \begin{pmatrix} n_y \cdot z - n_z \cdot r \cdot \sin\phi \\ n_z \cdot r \cdot \cos\phi - n_x \cdot z \\ n_x \cdot r \cdot \sin\phi - n_y \cdot r \cdot \cos\phi \end{pmatrix} \tag{4.120}$$

Transformation of a vector from the Cartesian coordinate system to the cylindrical coordinate system [5]:

$$\vec{v}(r, \phi, z) = \begin{pmatrix} v_x \cdot \cos\phi + v_y \cdot \sin\phi \\ -v_x \cdot \sin\phi + v_y \cdot \cos\phi \\ v_z \end{pmatrix} \tag{4.121}$$

From this, for this problem, it follows that

$$\vec{v} = \omega \cdot \begin{pmatrix} \left(n_y \cdot z - n_z \cdot r \cdot \sin\phi\right) \cdot \cos\phi + (n_z \cdot r \cdot \cos\phi - n_x \cdot z) \cdot \sin\phi \\ -\left(n_y \cdot z - n_z \cdot r \cdot \sin\phi\right) \cdot \sin\phi + (n_z \cdot r \cdot \cos\phi - n_x \cdot z) \cdot \cos\phi \\ n_x \cdot r \cdot \sin\phi - n_y \cdot r \cdot \cos\phi \end{pmatrix} \tag{4.122}$$

$$\vec{v} = \omega \cdot \begin{pmatrix} n_y \cdot z \cdot \cos\phi - n_z \cdot r \cdot \sin\phi \cdot \cos\phi + n_z \cdot r \cdot \cos\phi \cdot \sin\phi - n_x \cdot z \cdot \sin\phi \\ -n_y \cdot z \cdot \sin\phi + n_z \cdot r \cdot \sin\phi \cdot \sin\phi + n_z \cdot r \cdot \cos\phi \cdot \cos\phi - n_x \cdot z \cdot \cos\phi \\ n_x \cdot r \cdot \sin\phi - n_y \cdot r \cdot \cos\phi \end{pmatrix} \tag{4.123}$$

$$\vec{v} = \omega \cdot \begin{pmatrix} n_y \cdot z \cdot \cos\phi - n_x \cdot z \cdot \sin\phi \\ -n_y \cdot z \cdot \sin\phi - n_x \cdot z \cdot \cos\phi + n_z \cdot r \\ n_x \cdot r \cdot \sin\phi - n_y \cdot r \cdot \cos\phi \end{pmatrix} \tag{4.124}$$

$$rot\,\vec{v} = \omega \cdot rot \begin{pmatrix} n_y \cdot z \cdot \cos\phi - n_x \cdot z \cdot \sin\phi \\ -n_y \cdot z \cdot \sin\phi - n_x \cdot z \cdot \cos\phi + n_z \cdot r \\ n_x \cdot r \cdot \sin\phi - n_y \cdot r \cdot \cos\phi \end{pmatrix} \tag{4.125}$$

$$rot\,\vec{u} = \begin{pmatrix} \frac{1}{r} \cdot \frac{du_z}{d\phi} - \frac{du_\phi}{dz} \\ \frac{du_r}{dz} - \frac{du_z}{dr} \\ \frac{1}{r} \cdot \frac{d(r \cdot u_\phi)}{dr} - \frac{1}{r} \cdot \frac{du_r}{d\phi} \end{pmatrix} \cdot \begin{pmatrix} \vec{e_r} \\ \vec{e_\phi} \\ \vec{e_z} \end{pmatrix} \tag{4.126}$$

$$rot\,\vec{v} = \omega \cdot \begin{pmatrix} \frac{1}{r} \cdot \frac{d}{d\phi} \cdot (n_x \cdot r \cdot \sin\phi - n_y \cdot r \cdot \cos\phi) - \frac{d}{dz} \cdot (-n_y \cdot z \cdot \sin\phi - n_x \cdot z \cdot \cos\phi + n_z \cdot r) \\ \frac{d}{dz} \cdot (n_y \cdot z \cdot \cos\phi - n_x \cdot z \cdot \sin\phi) - \frac{d}{dr} \cdot (n_x \cdot r \cdot \sin\phi - n_y \cdot r \cdot \cos\phi) \\ \frac{1}{r} \cdot \frac{d(r \cdot (-n_y \cdot z \cdot \sin\phi - n_x \cdot z \cdot \cos\phi + n_z \cdot r))}{dr} - \frac{1}{r} \cdot \frac{d}{d\phi} \cdot (n_y \cdot z \cdot \cos\phi - n_x \cdot z \cdot \sin\phi) \end{pmatrix} \tag{4.127}$$

$$rot\,\vec{v} = \omega \cdot \begin{pmatrix} \frac{1}{r} \cdot (n_x \cdot r \cdot \cos\phi + n_y \cdot r \cdot \sin\phi) - (-n_y \cdot \sin\phi - n_x \cdot \cos\phi) \\ (n_y \cdot \cos\phi - n_x \cdot \sin\phi) - (n_x \cdot \sin\phi - n_y \cdot \cos\phi) \\ \frac{1}{r} \cdot (-n_y \cdot z \cdot \sin\phi - n_x \cdot z \cdot \cos\phi + 2 \cdot n_z \cdot r) - \frac{1}{r} \cdot (-n_y \cdot z \cdot \sin\phi - n_x \cdot z \cdot \cos\phi) \end{pmatrix} \tag{4.128}$$

$$rot\,\vec{v} = 2 \cdot \omega \cdot \begin{pmatrix} n_x \cdot \cos\phi + n_y \cdot \sin\phi \\ -n_x \cdot \sin\phi + n_y \cdot \cos\phi \\ \frac{n_z \cdot r}{r} \end{pmatrix} \tag{4.129}$$

Transformation of a vector from cylindrical coordinates to Cartesian coordinates [5]:

$$\vec{v}(x, y, z) = \begin{pmatrix} v_x \\ v_y \\ v_z \end{pmatrix} = \begin{pmatrix} v_r \cdot \cos\phi - v_\phi \cdot \sin\phi \\ v_r \cdot \sin\phi + v_\phi \cdot \cos\phi \\ v_z \end{pmatrix} \tag{4.130}$$

From this, for this task, the back transformation of the vector into the Cartesian coordinate system yields

$$rot\,\vec{v} = 2 \cdot \omega \cdot \begin{pmatrix} (n_x \cdot \cos\phi + n_y \cdot \sin\phi) \cdot \cos\phi - (-n_x \cdot \sin\phi + n_y \cdot \cos\phi) \cdot \sin\phi \\ (n_x \cdot \cos\phi + n_y \cdot \sin\phi) \cdot \sin\phi + (-n_x \cdot \sin\phi + n_y \cdot \cos\phi) \cdot \cos\phi \\ \frac{n_z \cdot r}{r} \end{pmatrix} \tag{4.131}$$

$$rot\overrightarrow{v} = 2 \cdot \omega \cdot \begin{pmatrix} n_x \\ n_y \\ n_z \end{pmatrix} \tag{4.132}$$

$$rot\overrightarrow{v} = 2 \cdot \omega \cdot \overrightarrow{n} \tag{4.133}$$

q.e.d. [5]

This means that as long as the vector field r is confined to a plane parallel to that spanned by the x- and y-axes (z is independent of the parameter φ), the rotation function described in [5] is valid.

If the vector field r is extended such that the plane on which the vector field lies is tilted by the angle α about the x-axis, the following occurs:

$$\overrightarrow{v} = \omega \cdot \left(\overrightarrow{n} \times \overrightarrow{r}\right) \tag{4.134}$$

With

$$\overrightarrow{n} = \begin{pmatrix} n_x \\ n_y \\ n_z \end{pmatrix} \tag{4.135}$$

and

$$\overrightarrow{r} = \begin{pmatrix} r \cdot \cos\phi \\ r \cdot \cos\alpha \cdot \sin\phi \\ r \cdot \sin\alpha \cdot \sin\phi \end{pmatrix} \tag{4.136}$$

this yields

$$\overrightarrow{v} = \omega \cdot \left(\begin{pmatrix} n_x \\ n_y \\ n_z \end{pmatrix} \times \begin{pmatrix} r \cdot \cos\phi \\ r \cdot \cos\alpha \cdot \sin\phi \\ r \cdot \sin\alpha \cdot \sin\phi \end{pmatrix} \right) \tag{4.137}$$

$$\overrightarrow{v} = \omega \cdot \begin{pmatrix} n_y \cdot r \cdot \sin\alpha \cdot \sin\phi - n_z \cdot r \cdot \cos\alpha \cdot \sin\phi \\ n_z \cdot r \cdot \cos\phi - n_x \cdot r \cdot \sin\alpha \cdot \sin\phi \\ n_x \cdot r \cdot \cos\alpha \cdot \sin\phi - n_y \cdot r \cdot \cos\phi \end{pmatrix} \tag{4.138}$$

Transformation into the cylindrical coordinate system with

$$\overrightarrow{v}(r,\phi,z) = \begin{pmatrix} v_x \cdot \cos\phi + v_y \cdot \sin\phi \\ -v_x \cdot \sin\phi + v_y \cdot \cos\phi \\ v_z \end{pmatrix} \tag{4.139}$$

yields

$$\vec{v}(r,\phi,z) = \omega \cdot \begin{pmatrix} (n_y \cdot r \cdot \sin\alpha \cdot \sin\phi - n_z \cdot r \cdot \cos\alpha \cdot \sin\phi) \cdot \cos\phi + (n_z \cdot r \cdot \cos\phi - n_x \cdot r \cdot \sin\alpha \cdot \sin\phi) \cdot \sin\phi \\ -(n_y \cdot r \cdot \sin\alpha \cdot \sin\phi - n_z \cdot r \cdot \cos\alpha \cdot \sin\phi) \cdot \sin\phi + (n_z \cdot r \cdot \cos\phi - n_x \cdot r \cdot \sin\alpha \cdot \sin\phi) \cdot \cos\phi \\ n_x \cdot r \cdot \cos\alpha \cdot \sin\phi - n_y \cdot r \cdot \cos\phi \end{pmatrix} \tag{4.140}$$

$$\vec{v}(r,\phi,z) = \omega \cdot \begin{pmatrix} -n_x \cdot r \cdot \sin\alpha \cdot \sin\phi \cdot \sin\phi + n_y \cdot r \cdot \sin\alpha \cdot \sin\phi \cdot \cos\phi \\ -n_z \cdot r \cdot \cos\alpha \cdot \sin\phi \cdot \cos\phi + n_z \cdot r \cdot \cos\phi \cdot \sin\phi \\ -n_x \cdot r \cdot \sin\alpha \cdot \sin\phi \cdot \cos\phi - n_y \cdot r \cdot \sin\alpha \cdot \sin\phi \cdot \sin\phi \\ +n_z \cdot r \cdot \cos\alpha \cdot \sin\phi \cdot \sin\phi + n_z \cdot r \cdot \cos\phi \cdot \cos\phi \\ n_x \cdot r \cdot \cos\alpha \cdot \sin\phi - n_y \cdot r \cdot \cos\phi \end{pmatrix} \tag{4.141}$$

Applying the rotation function

$$rot\,\vec{v}(r,\phi,z) = \omega \cdot rot \begin{pmatrix} -n_x \cdot r \cdot \sin\alpha \cdot \sin\phi \cdot \sin\phi + n_y \cdot r \cdot \sin\alpha \cdot \sin\phi \cdot \cos\phi \\ -n_z \cdot r \cdot \cos\alpha \cdot \sin\phi \cdot \cos\phi + n_z \cdot r \cdot \cos\phi \cdot \sin\phi \\ -n_x \cdot r \cdot \sin\alpha \cdot \sin\phi \cdot \cos\phi - n_y \cdot r \cdot \sin\alpha \cdot \sin\phi \cdot \sin\phi \\ +n_z \cdot r \cdot \cos\alpha \cdot \sin\phi \cdot \sin\phi + n_z \cdot r \cdot \cos\phi \cdot \cos\phi \\ n_x \cdot r \cdot \cos\alpha \cdot \sin\phi - n_y \cdot r \cdot \cos\phi \end{pmatrix} \tag{4.142}$$

$$rot\,\vec{v}(r,\phi,z) = \omega \cdot \begin{pmatrix} \frac{1}{r} \cdot \frac{d}{d\phi} \cdot (n_x \cdot r \cdot \cos\alpha \cdot \sin\phi - n_y \cdot r \cdot \cos\phi) \\ -\frac{d}{dz} \cdot \begin{pmatrix} -n_x \cdot r \cdot \sin\alpha \cdot \sin\phi \cdot \cos\phi - n_y \cdot r \cdot \sin\alpha \cdot \sin\phi \cdot \sin\phi \\ +n_z \cdot r \cdot \cos\alpha \cdot \sin\phi \cdot \sin\phi + n_z \cdot r \cdot \cos\phi \cdot \cos\phi \end{pmatrix} \\ \frac{d}{dz} \cdot \begin{pmatrix} -n_x \cdot r \cdot \sin\alpha \cdot \sin\phi \cdot \sin\phi + n_y \cdot r \cdot \sin\alpha \cdot \sin\phi \cdot \cos\phi \\ -n_z \cdot r \cdot \cos\alpha \cdot \sin\phi \cdot \cos\phi + n_z \cdot r \cdot \cos\phi \cdot \sin\phi \end{pmatrix} \\ -\frac{d}{dr} \cdot (n_x \cdot r \cdot \cos\alpha \cdot \sin\phi - n_y \cdot r \cdot \cos\phi) \\ \frac{1}{r} \cdot \frac{d\left(r \cdot \begin{pmatrix} -n_x \cdot r \cdot \sin\alpha \cdot \sin\phi \cdot \cos\phi - n_y \cdot r \cdot \sin\alpha \cdot \sin\phi \cdot \sin\phi \\ +n_z \cdot r \cdot \cos\alpha \cdot \sin\phi \cdot \sin\phi + n_z \cdot r \cdot \cos\phi \cdot \cos\phi \end{pmatrix}\right)}{dr} \\ -\frac{1}{r} \cdot \frac{d}{d\phi} \cdot \begin{pmatrix} -n_x \cdot r \cdot \sin\alpha \cdot \sin\phi \cdot \sin\phi + n_y \cdot r \cdot \sin\alpha \cdot \sin\phi \cdot \cos\phi \\ -n_z \cdot r \cdot \cos\alpha \cdot \sin\phi \cdot \cos\phi + n_z \cdot r \cdot \cos\phi \cdot \sin\phi \end{pmatrix} \end{pmatrix} \tag{4.143}$$

$$\xrightarrow{\hspace{20em}}$$

$$rot\ \vec{v}\ (r,\varphi,z) = \omega \cdot \begin{pmatrix} n_x \cdot \cos\alpha \cdot \cos\varphi + n_y \cdot \sin\varphi \\ -n_x \cdot \cos\alpha \cdot \sin\varphi + n_y \cdot \cos\varphi \\ -2 \cdot n_x \cdot \sin\alpha \cdot \sin\varphi \cdot \cos\varphi - 2 \cdot n_y \cdot \sin\alpha \cdot \sin\varphi \cdot \sin\varphi \\ +2 \cdot n_z \cdot \cos\alpha \cdot \sin\varphi \cdot \sin\varphi + 2 \cdot n_z \cdot \cos\varphi \cdot \cos\varphi \\ -\frac{d}{d\varphi} \cdot \begin{pmatrix} -n_x \cdot \sin\alpha \cdot \sin\varphi \cdot \sin\varphi + n_y \cdot \sin\alpha \cdot \sin\varphi \cdot \cos\varphi \\ -n_z \cdot \cos\alpha \cdot \sin\varphi \cdot \cos\varphi + n_z \cdot \cos\varphi \cdot \sin\varphi \end{pmatrix} \end{pmatrix} \tag{4.144}$$

Auxiliary calculation or [5]:

$$\frac{d}{d\phi}(\sin\phi \cdot \sin\phi) = 2 \cdot \cos\phi \cdot \sin\phi \tag{4.145}$$

$$\frac{d}{d\phi}(\sin\phi \cdot \cos\phi) = (\cos\phi \cdot \cos\phi - \sin\phi \cdot \sin\phi) \tag{4.146}$$

End of auxiliary calculation

$$rot\vec{v}\,(r,\varphi,z)\omega\cdot \begin{pmatrix} n_x \cdot \cos\alpha \cdot \cos\varphi + n_y \cdot \sin\varphi \\ -n_x \cdot \cos\alpha \cdot \sin\varphi + n_y \cdot \cos\varphi \\ -2 \cdot n_x \cdot \sin\alpha \cdot \sin\varphi \cdot \cos\varphi - 2 \cdot n_y \cdot \sin\alpha \cdot \sin\varphi \cdot \sin\varphi \\ +2 \cdot n_z \cdot \cos\alpha \cdot \sin\varphi \cdot \sin\varphi + 2 \cdot n_z \cdot \cos\varphi \cdot \cos\varphi \\ -\begin{pmatrix} -n_x \cdot \sin\alpha \cdot 2 \cdot \cos\varphi \cdot \sin\varphi + n_y \cdot \sin\alpha \cdot (\cos\varphi \cdot \cos\varphi - \sin\varphi \cdot \sin\varphi) \\ -n_z \cdot \cos\alpha \cdot (\cos\varphi \cdot \cos\varphi - \sin\varphi \cdot \sin\varphi) + n_z \cdot (\cos\varphi \cdot \cos\varphi - \sin\varphi \cdot \sin\varphi) \end{pmatrix} \end{pmatrix} \tag{4.147}$$

$$rot\ \vec{v}\ (r,\varphi,z) = \omega \cdot \begin{pmatrix} n_x \cdot \cos\alpha \cdot \cos\varphi + n_y \cdot \sin\varphi \\ -n_x \cdot \cos\alpha \cdot \sin\varphi + n_y \cdot \cos\varphi \\ -2 \cdot n_x \cdot \sin\alpha \cdot \sin\varphi \cdot \cos\varphi - 2 \cdot n_y \cdot \sin\alpha \cdot \sin\varphi \cdot \sin\varphi \\ +2 \cdot n_z \cdot \cos\alpha \cdot \sin\varphi \cdot \sin\varphi + 2 \cdot n_z \cdot \cos\varphi \cdot \cos\varphi \\ +n_x \cdot \sin\alpha \cdot 2 \cdot \cos\varphi \cdot \sin\varphi \\ -n_y \cdot \sin\alpha \cdot \cos\varphi \cdot \cos\varphi + n_y \cdot \sin\alpha \cdot \sin\varphi \cdot \sin\varphi \\ +n_z \cdot \cos\alpha \cdot \cos\varphi \cdot \cos\varphi - n_z \cdot \cos\alpha \cdot \sin\varphi \cdot \sin\varphi \\ -n_z \cdot \cos\varphi \cdot \cos\varphi + n_z \cdot \sin\varphi \cdot \sin\varphi \end{pmatrix} \tag{4.148}$$

$$rot\ \vec{v}\,(r,\varphi,z) = \omega \cdot \begin{pmatrix} n_x \cdot \cos\alpha \cdot \cos\varphi + n_y \cdot \sin\varphi \\ -n_x \cdot \cos\alpha \cdot \sin\varphi + n_y \cdot \cos\varphi \\ -2 \cdot n_x \cdot \sin\alpha \cdot \sin\varphi \cdot \cos\varphi + 2 \cdot n_x \cdot \sin\alpha \cdot \cos\varphi \cdot \sin\varphi \\ -2 \cdot n_y \cdot \sin\alpha \cdot \sin\varphi \cdot \sin\varphi \\ -n_y \cdot \sin\alpha \cdot \cos\varphi \cdot \cos\varphi + n_y \cdot \sin\alpha \cdot \sin\varphi \cdot \sin\varphi \\ +2 \cdot n_z \cdot \cos\alpha \cdot \sin\varphi \cdot \sin\varphi + 2 \cdot n_z \cdot \cos\varphi \cdot \cos\varphi \\ +n_z \cdot \cos\alpha \cdot \cos\varphi \cdot \cos\varphi - n_z \cdot \cos\alpha \cdot \sin\varphi \cdot \sin\varphi \\ -n_z \cdot \cos\varphi \cdot \cos\varphi + n_z \cdot \sin\varphi \cdot \sin\varphi \end{pmatrix} \tag{4.149}$$

$$rot\,\vec{v}\,(r,\phi,z) = \omega \cdot \begin{pmatrix} n_x \cdot \cos\alpha \cdot \cos\phi + n_y \cdot \sin\phi \\ -n_x \cdot \cos\alpha \cdot \sin\phi + n_y \cdot \cos\phi \\ -n_y \cdot \sin\alpha + n_z \cdot \cos\alpha + n_z \end{pmatrix} \tag{4.150}$$

Back transformation to Cartesian coordinates:

$$rot\,\vec{v}\,(x,y,z) = \omega \cdot \begin{pmatrix} \left(n_x \cdot \cos\alpha \cdot \cos\phi + n_y \cdot \sin\phi\right) \cdot \cos\phi \\ -\left(-n_x \cdot \cos\alpha \cdot \sin\phi + n_y \cdot \cos\phi\right) \cdot \sin\phi \\ \left(n_x \cdot \cos\alpha \cdot \cos\phi + n_y \cdot \sin\phi\right) \cdot \sin\phi \\ +\left(-n_x \cdot \cos\alpha \cdot \sin\phi + n_y \cdot \cos\phi\right) \cdot \cos\phi \\ -n_y \cdot \sin\alpha + n_z \cdot \cos\alpha + n_z \end{pmatrix} \tag{4.151}$$

$$rot\,\vec{v}\,(x,y,z) = \omega \cdot \begin{pmatrix} n_x \cdot \cos\alpha \cdot \cos\phi \cdot \cos\phi + n_y \cdot \sin\phi \cdot \cos\phi \\ +n_x \cdot \cos\alpha \cdot \sin\phi \cdot \sin\phi - n_y \cdot \cos\phi \cdot \sin\phi \\ n_x \cdot \cos\alpha \cdot \cos\phi \cdot \sin\phi + n_y \cdot \sin\phi \cdot \sin\phi \\ -n_x \cdot \cos\alpha \cdot \sin\phi \cdot \cos\phi + n_y \cdot \cos\phi \cdot \cos\phi \\ -n_y \cdot \sin\alpha + n_z \cdot \cos\alpha + n_z \end{pmatrix} \tag{4.152}$$

$$rot\,\vec{v}\,(x,y,z) = \omega \cdot \begin{pmatrix} n_x \cdot \cos\alpha \\ n_y \\ -n_y \cdot \sin\alpha + n_z \cdot \cos\alpha + n_z \end{pmatrix} \tag{4.153}$$

For $\alpha = 0$, the result is

$$rot\,\vec{v}\,(x,y,z) = \omega \cdot \begin{pmatrix} n_x \\ n_y \\ 2 \cdot n_z \end{pmatrix} \tag{4.154}$$

Since $n_x = n_y = 0$, it follows that

$$rot\,\vec{v}\,(x, y, z) = 2 \cdot \omega \cdot \begin{pmatrix} n_x \\ n_y \\ n_z \end{pmatrix} \tag{4.155}$$

The result corresponds to what was derived above.

For $\alpha = 90°$, the result is

$$rot\,\vec{v}\,(x, y, z) = \omega \cdot \begin{pmatrix} 0 \\ n_y \\ -n_y + n_z \end{pmatrix} \tag{4.156}$$

Since the plane is tilted about the x-axis, $n_x = n_z = 0$.

Thus,

$$rot\,\vec{v}\,(x, y, z) = \omega \cdot \begin{pmatrix} 0 \\ n_y \\ -n_y \end{pmatrix} \tag{4.157}$$

This is not

$$rot\,\vec{v}\,(x, y, z) = 2 \cdot \omega \cdot \begin{pmatrix} n_x \\ n_y \\ n_z \end{pmatrix} \tag{4.158}$$

It follows that the rotation function can only be applied if the vector field is a planar vector field in the x–y plane. A corresponding note can also be found in [6].

A meaningful application of the rotation function in (more complex) three-dimensional fields, as described by the toroidal helix, is therefore not possible.

However, the fundamental insight from the rotation function in the differential form of Maxwell's equations—that the electric and magnetic fields are perpendicular to each other and circulate around one another—can be used to describe the functions in the toroidal helix and, among other things, to derive the cause of the oscillations.

One possible reason for the limited applicability of the rotation function may be that the coordinate transformation between the cylindrical and Cartesian coordinate systems, as described by [5], is only valid to a limited extent (see Chap. 10 Coordinate Transformations).

4.2 The Photon Mass

Frequencies of electromagnetic origin occurring in nature can be divided into spectra as follows [7] (Table 4.1):

To address the question of the size of the helical structure as discussed in Section 4, the fundamental physical building blocks are considered [8]:

The spatial dimensions of the elements listed in the Standard Model fall within the following ranges:

Electron [9]: 10^{-19} m to point-like

Myon [10]: point-like

Tau: no information available

It can be assumed that the wavelength of the fundamental frequency at which the toroidal helices oscillate is at most in the range of 10^{-19} m, since the developing waves must be on the order of the diameter of the toroidal helix.

Furthermore, it can be assumed that the excitation frequency of the toroidal helices, the frequency for the lateral, and the frequency for the longitudinal fundamental oscillation are identical, where "longitudinal fundamental oscillation" means that the resulting field is the static electric field.

The rest mass of the electron is $9.10938356 \cdot 10^{-31}$ kg

The electric charge of the electron is $-1.6021766208 \cdot 10^{-19}$ C

If the two forces resulting from these quantities are compared, a factor is obtained that provides information about the magnitude difference between the two forces.

Two electrons repel each other on the one hand because they have the same electric charge; on the other hand, they attract each other because they have mass.

$$F_e = \frac{q_1 \cdot q_2}{4 \cdot \pi \cdot \varepsilon_0 \cdot \varepsilon_r \cdot r^2} \tag{4.159}$$

$$F_g = \frac{g \cdot m_1 \cdot m_2}{r^2} \tag{4.160}$$

From these two forces, the ratio a is now determined in order to obtain the ratio for the geometric dimensions of the toroidal helix with respect to the diameter of the ring and the diameter of the toroidal body.

$$a = \frac{F_e}{F_g} \tag{4.161}$$

$$a = \frac{\frac{q_1 \cdot q_2}{4 \cdot \pi \cdot \varepsilon_0 \cdot \varepsilon_r \cdot r^2}}{\frac{g \cdot m_1 \cdot m_2}{r^2}} \tag{4.162}$$

$$a = \frac{q_1 \cdot q_2}{4 \cdot \pi \cdot \varepsilon_0 \cdot \varepsilon_r \cdot g \cdot m_1 \cdot m_2} \tag{4.163}$$

Tab. 4.1 Electromagnetic Spectrum

Name of frequency range	Subdivision	Wavelength		Frequency		Photon energy	Generation/excitation	Technical application
		from	to	from	to			
Low frequency	Extremely Low Frequency (ELF)	10 Mm	100 Mm	3 Hz	30 Hz	$>2.0\times10^{-33}$ J >12 feV	Earth dipole, antenna systems	Railway power
	Super Low Frequency (SLF)	1 Mm	10 Mm	30 Hz	300 Hz	$>2.0\times10^{-32}$ J >120 feV		Mains frequency, former submarine communication
	Ultra Low Frequency (ULF)	100 km	1000 km	300 Hz	3 kHz	$>2.0\times10^{-31}$ J >1.2 peV		
	Very Low Frequency (VLF), long waves (SLW)	10 km	100 km	3 kHz	30 kHz	$>2.0\times10^{-30}$ J >12 peV		Submarine communication, radio navigation, pulse clocks
Radio waves	Long wave (LW)	1 km	10 km	30 kHz	300 kHz	$>2.0\times10^{-29}$ J >120 peV	Oscillator circuit + antenna	Long wave RF, DCF 77
	Medium wave (MW)	100 m	1000 m	300 kHz	3 MHz	$>2.0\times10^{-28}$ J >1.2 neV		Medium wave RF, RF surgery, boundary wave, shortwave RF
	Short wave (SW)	10 m	100 m	3 MHz	30 MHz	$>2.0\times10^{-27}$ J >12 neV		Boundary wave, shortwave RF, diathermy, RC model building
	Very high frequency (VHF)	1 m	10 m	30 MHz	300 MHz	$>2.0\times10^{-26}$ J >120 neV	Nuclear magnetic resonance excitation	Radio, television, radar, magnetic resonance imaging

(continued)

Tab. 4.1 (continued)

Name of frequency range	Subdivision	Wavelength		Frequency		Photon energy	Generation/excitation	Technical application
		from	to	from	to			
Microwaves	Decimeter waves	10 cm	1 m	300 MHz	3 GHz	>2.0×10^{-25} J >1.2 μeV	Magnetron, klystron, maser, cosmic background radiation	Radar, magnetic resonance imaging, mobile communications, television, microwave oven, WLAN, Bluetooth, GPS
	Centimeter waves	1 cm	10 cm	3 GHz	30 GHz	>2.0×10^{-24} J >12 μeV	Excitation of nuclear magnetic resonance and electron spin resonance, molecular rotations	Radar, radio astronomy, directional radio, satellite broadcasting, WLAN
	Millimeter waves	1 mm	1 cm	30 GHz	300 GHz	> 2.0×10^{-23} J >120 μeV		Radar, radio astronomy, directional radio
Terahertz radiation		30 μm	3 mm	0.1 THz	10 THz	>6.6×10^{-23} J >0.4 meV	Synchrotron, free-electron laser	Radio astronomy, spectroscopy, imaging techniques, security technology
Infrared radiation (thermal radiation)	Far infrared	50 μm	1 mm	300 GHz	6 THz	> 2.0×10^{-22} J > 1.2 meV	Thermal radiator, synchrotron	Infrared spectroscopy, Raman spectroscopy, infrared astronomy
	Mid infrared	3.0 μm	50 μm	6 THz	100 THz	>4.0×10^{-21} J >25 meV	Carbon dioxide laser, quantum cascade laser	Thermography
	Near infrared	780 nm	3.0 μm	100 THz	385 THz	>8.0×10^{-20} J > 500 meV	Nd:YAG laser, laser diode	Remote control, data communication (IRDA), CD

(continued)

Tab. 4.1 (continued)

Name of frequency range	Subdivision	Wavelength		Frequency		Photon energy	Generation/excitation	Technical application
		from	to	from	to			
Light	Red	640 nm	780 nm	384 THz	468 THz	1.6–1.95 eV	Thermal radiator (incandescent lamp), gas discharge (fluorescent tube), dye and other lasers, synchrotron	DVD, laser pointer, red, green: laser level, lighting, colorimetry, photometry, red, yellow, green: traffic signal, violet: Blu-ray Disc
	Orange	600 nm	640 nm	468 THz	500 THz	1.95–2.06 eV		
	Yellow	570 nm	600 nm	500 THz	526 THz	2.06–2.17 eV		
	Green	490 nm	570 nm	526 THz	612 THz	2.17–2.53 eV		
	Blue	430 nm	490 nm	612 THz	697 THz	2.53–2.88 eV		
	Violet	380 nm	430 nm	697 THz	789 THz	$>4.6\times10^{-19}$ J >2.9 eV	Excitation of valence electrons	
UV radiation	Weak UV radiation	200 nm	380 nm	789 THz	1.5 PHz	$>5.2\times10^{-19}$ J >3.3 eV	Gas discharge, synchrotron, excimer laser	Black light, fluorescence, phosphorescence, banknote verification, photolithography, disinfection, UV light, spectroscopy
	Strong UV radiation	50 nm	200 nm	1.5 PHz	6 PHz	$>9.9\times10^{-19}$ J >6.2 eV		
	XUV	1 nm	50 nm	6 PHz	300 PHz	$>5.0\times10^{-18}$ J 20–1000 eV	XUV tube, synchrotron, nanoplasma	EUV lithography, X-ray microscopy, nanoscopy
X-rays		10 pm	1 nm	300 PHz	30 EHz	$>2.0\times10^{-16}$ J >1 keV	X-ray tube, synchrotron, excitation of inner electrons, Auger electrons	medical diagnostics, security technology, X-ray structure analysis, X-ray diffraction, photoelectron spectroscopy, X-ray absorption spectroscopy

(continued)

Tab. 4.1 (continued)

Name of frequency range	Subdivision	Wavelength		Frequency		Photon energy	Generation/excitation	Technical application
		from	to	from	to			
Gamma rays		80 fm	10 pm	30 EHz	3.75 ZHz	$>2.0\times10^{-4}$ J >120 keV	Radioactivity, annihilation, excitation of nuclear states	medical radiation therapy, Mössbauer spectroscopy
(Cosmic rays)		1 fm	80 fm	3.75 ZHz	300 ZHz (10^{21})			

$$a = \frac{\left(-1.6021766208 \cdot 10^{-19} As\right)^2}{4 \cdot \pi \cdot 8.85418781762 \cdot 10^{-12} \frac{As}{Vm} \cdot 6.67408 \cdot 10^{-11} \frac{m^3}{kg \cdot s^2} \cdot \left(9.10938356 \cdot 10^{-31} kg\right)^2} \tag{4.164}$$

$$a = \frac{(-1.6021766208)^2}{4 \cdot \pi \cdot 8.85418781762 \cdot 6.67408 \cdot (9.10938356)^2} \cdot \frac{10^{-38}}{10^{-12} \cdot 10^{-11} \cdot 10^{-62}} \frac{A^2 \cdot s^2}{\frac{As}{Vm} \cdot \frac{m^3}{kg \cdot s^2} \cdot kg^2} \tag{4.165}$$

$$a = \frac{1.6021766208^2}{4 \cdot \pi \cdot 8.85418781762 \cdot 6,67408 \cdot 9.10938356^2} \cdot 10^{47} \frac{A^2 \cdot s^2 \cdot V \cdot m \cdot kg \cdot s^2}{A \cdot s \cdot m^3 \cdot kg^2} \tag{4.166}$$

$$a = 1.27760825512997 \cdot 10^{45} \frac{A \cdot V \cdot s^3}{m^2 \cdot kg} \tag{4.167}$$

Supplementary calculation:

Electrical power:

$$P_{el} = U \cdot I [V \cdot A] \tag{4.168}$$

Mechanical power:

$$P_{mech} = \frac{m \cdot a \cdot s}{t} \left[\frac{kg \cdot m \cdot m}{s^2 \cdot s} \right] \tag{4.169}$$

End of the supplementary calculation.

The two unit blocks for electrical and mechanical power cancel each other out.

$$a = 1.27760825512997 \cdot 10^{45} \tag{4.170}$$

This factor a represents the ratio of the forces that arise from the square of the two electromagnetic effects generated by the toroidal coil. The square is considered because, in both calculations, the multiplications of the masses on the one hand and the charges on the other appear ($m_1 \cdot m_2$ and $q_1 \cdot q_2$, respectively).

By taking the square root of a, the ratio of the two electromagnetic effects can now be determined:

$$v_{ef} = \sqrt{a} \tag{4.171}$$

$$v_{ef} = \sqrt{1.27760825512997 \cdot 10^{45}} \tag{4.172}$$

$$v_{ef} = \sqrt{12.7760825512997} \cdot \sqrt{10^{44}} \tag{4.173}$$

$$v_{ef} = 3.57436463602969 \cdot 10^{22} \tag{4.174}$$

This means that the mass ratio between the smallest possible gravitational element (photon, gluon) and the known electrical element, the electron, is at least in the range of 10^{22} (see Standard Model overview, Table 4.2). The mass of the photon is thus in the range of 10^{-52} … 10^{-53} kg. This estimate is based on the initial assumption that the gravitational force, which manifests as mass, is an electromagnetic effect.

Tab. 4.2 Standard Model of the fundamental building blocks of matter

		Three generations of matter (fermions)			Vector bosons	
		I	II	III		
Mass	Quarks	2.3 MeV	1.275 GeV	173.07 GeV	0	125.09 GeV
Charge		2/3	2/3	2/3	0	0
Spin		1/2	1/2	1/2	1	0
Symbol		u	c	t	Y	H
Name		up	charm	top	Photon	Higgs boson
Mass		4.8 MeV	95 MeV	4.18 GeV	0	
Charge		-1/3	-1/3	-1/3	0	
Spin		1/2	1/2	1/2	1	
Symbol		d	s	b	g	
Name		down	strange	bottom	Gluon	
Mass	Leptons	< 2 eV	< 0.19 eV	< 18.2 MeV	91.2 GeV	
Charge		0	0	0	0	
Spin		1/2	1/2	1/2	1	
Symbol		ν_e	ν_μ	ν_τ	Z^0	
Name		electron Neutrino	muon Neutrino	Tau-Neutrino	Z-Boson	
Mass		0.511 MeV	105.7 MeV	1.777 GeV	80.4 GeV	
Charge		-1	-1	-1	±1	
Spin		1/2	1/2	1/2	1	
Symbol		e	μ	τ	$W^\pm$	
Name		Electron	Muon	Tau	W-Boson	

If one considers the calculation of the deviation angle γ from Eq. 4.98 or the maximum values for the angle as a function of the angle α, the following relationship can be observed:

$$\gamma = \arccos \frac{2 \cdot a^2}{\sqrt{\cos^4(a \cdot \alpha) + 4 \cdot a^4}} \tag{4.175}$$

The angle γ reaches its maximum value when the argument for the arccos function is at its minimum, because the arccos function for the principal value is a monotonically decreasing function.

As a side note: The variable a in Eq. 4.171 has nothing to do with the variable a in Eq. 4.167.

The minimum value for the argument occurs when the denominator of the fraction is at its maximum. This is the case when cos (a · α) takes the value 1, which happens when $\alpha = 0$.

Thus,

$$\gamma_{\max}(a) = \arccos \frac{2 \cdot a^2}{\sqrt{1 + 4 \cdot a^4}} \tag{4.176}$$

$\gamma_{max}(a)$ is approximately proportional to the reciprocal of the square of a.

Accordingly, the amplitude of the lateral oscillation, which causes the gravitational behavior, is also proportional to the reciprocal of the square of a. At the same time, the oscillation amplitude of the longitudinal oscillation, which causes the electrical behavior, is proportional to the reciprocal of a.

It follows that if a from Eq. 4.170 is in the range of 10^{22}, then the amplitude of the longitudinal oscillation is greater by a factor of 10^{22} than that of the lateral oscillation, and thus the forces caused by the oscillation are greater by this factor of 10^{22}.

Furthermore, this means that if the radius R for the toroid is about 10^{-19} m (the size of an electron), then the radius r for the toroidal cross-section is about 10^{-41} m.

Assuming that the wavelength of the natural resonance is in the range of $r = 10^{-41}$ m, then the frequency is about 10^{49} Hz according to the equation

$$\lambda = \frac{c}{f}. \tag{4.177}$$

This frequency should be measurable when the gravitational field is investigated. Furthermore, this frequency must be an integer multiple of all photon frequencies. This conclusion implies that the photon spectrum is not continuous, but discrete, i.e., that there are only finitely many different frequencies in the photon spectrum, all of which have a least common multiple.

In summary, the toroidal coil thus represents the fundamental building block from which all elements of the Standard Model can be constructed and with which the four fundamental forces of physics can be explained.

4.3 The Strong Force

The strong force, or strong interaction, is one of the four fundamental forces in physics and is characterized by its short range. Due to the strong force, heavy particles such as protons and neutrons are held together at the subatomic level [11]. The strong force is mediated by gluons, which interact between the heavy particles and move among them.

4.4 The Weak Force

The weak force, or weak interaction, is one of the four fundamental forces in physics [12]. The weak force, with its extremely short range, becomes apparent, for example, when a proton and an electron are converted into a neutron with the release of energy. The weak interaction is mediated by the exchange of photons.

4.5 The Toroidal Helix and the Fundamental Elements

The toroidal helix consists of two interwoven and intertwined fields. These two fields are sustained by oscillating at the previously estimated frequency, thereby mutually inducing each other.

This oscillation can manifest as a standing wave on the toroid, or as a wave that propagates longitudinally and, accordingly, winds helically around the toroid. This outwardly expresses the spin of the toroid.

When several toroids are combined, they form a toroid bundle, resulting in a corresponding multiplication of the smallest gravitational field strength and, consequently, an increase in the mass of the resulting element, which also assumes a toroidal shape.

If a critical number of toroids come together to form a toroid bundle, a longitudinal oscillation of the magnetic helix may occur, i.e., an oscillation in the direction of the carrier circle of the toroid bundle, which externally results in an electric field that is, on average, either positively or negatively polarized.

Gluons are toroidal helices that act as phase equalizers between toroid bundles.

Photons are toroidal helices that are ejected from toroid bundles by a collision and are thereby excited to oscillate laterally, i.e., radially to the carrier circle of the toroid. The oscillation frequency is an integer fraction of the natural frequency on the toroid and carries the energy of the photon, which is released as energy upon impact.

4.6 The Toroidal Helix and the Quantum of Action

If one attempts to reconcile the preceding findings with Planck's quantum of action, such an attempt is doomed to fail [13]. In particular, if one

$$E = h \cdot f \tag{4.178}$$

with

E = energy in J
h = Planck's quantum of action = $6.62607015 \cdot 10^{-34}$ J s
f = frequency in Hz
is taken into account and

$$E = m \cdot c^2 \tag{4.179}$$

with

E = energy in J
m = mass in kg
c = speed of light in vacuum = 299.792.458 m/s
are equated:

$$h \cdot f = m \cdot c^2 \tag{4.180}$$

By rearranging, one obtains

$$f = m \cdot \frac{c^2}{h} \tag{4.181}$$

$$f = m \cdot \frac{\left(2.99792458 \cdot 10^8 \frac{m}{s}\right)^2}{6.62607015 \cdot 10^{-34} Js} \tag{4.182}$$

$$f = m \cdot \frac{\left(2.99792458 \cdot 10^8 \frac{m}{s}\right)^2}{6.62607015 \cdot 10^{-34} Js} \tag{4.183}$$

$$f = m \cdot \frac{8.9875517873681764 \cdot 10^{16} \frac{m^2}{s^2}}{6.62607015 \cdot 10^{-34} Js} \tag{4.184}$$

$$f = m \cdot 1.35639249 \cdot 10^{50} \frac{kg \cdot m^2}{\frac{kg \cdot m \cdot m}{s^2} \cdot s^3} \tag{4.185}$$

$$f = m \cdot 1.35639249 \cdot 10^{50} \frac{1}{s} \tag{4.186}$$

If one compares this ratio with the one determined above (10^{-52} kg vs. 10^{49} Hz), a discrepancy of about 10^{51} becomes apparent.

A possible explanation for this discrepancy, in my view—and this gives the concept of wave-particle duality a new dimension—is that, in propagating light, the observable wavelength of light is not the fundamental frequency of the individual photon, but rather, similar to a traveling water wave, the phase velocity of a collection of photons, i.e., the frequency of the physical motion of the light quanta. This model is supported by the interference patterns observed in light phenomena at a slit, which correspond to those of water waves at a slit [14; 15].

References

1. Joseph Polchinski; String Theory – Volume I An Introduction to the Bosonic String; Cambridge University Press, Edition of 2007
2. Joseph Polchinski; String Theory – Volume II Superstring Theory and Beyond; Cambridge University Press; Edition of 2007
3. B. Marx, R. Süsse; Theoretische Elektrotechnik – Band 1: Variationsrechnung und Maxwellsche Gleichungen; BI Wissenschaftsverlag; 1994
4. Wikipedia, Maxwell-Gleichungen, 1. 3. 2021
5. Bronstein, Semendjajew; Taschenbuch der Mathematik; Verlag Nauka Moskau, 25. Auflage 1991
6. Klemens Burg, Herbert Haf, Friedrich Wille; Vektoranalysis – Höhere Mathematik für Ingenieure, Naturwissenschaftler und Mathematiker; Teubner Verlag Wiesbaden; 2006
7. Wikipedia, Elektromagnetisches Spektrum, 1. 3. 2021
8. www.teilchenphysik.at/wissen/das-standardmodell/, 1. 3. 2021
9. Wikipedia, Elektron, 1. 3. 2021
10. Wikipedia, Myon, 1. 3. 2021
11. Wikipedia, Starke Wechselwirkung, 1. 3. 2021
12. Wikipedia, Schwache Wechselwirkung, 1. 3. 2021
13. Wikipedia, Plancksches Wirkungsquantum, 1. 3. 2021
14. Wikipedia, Interferenz, 1. 3. 2021
15. Jay O'Rear; Physik; Carl Hanser Verlag München; 1989

5 Relativity

The theories of relativity deal with the behavior of photons, which move at a finite speed of light and, depending on the observer's position, lead to seemingly paradoxical relationships [1; 2]. Upon closer examination, however, these behaviors relative to the observer's position are understandable, as is the fact that photons always travel at the speed of light.

However, the approaches derived from the observations that photons have no mass and that mass cannot be accelerated to the speed of light are bold, since accelerated mass increases by a factor of

$$\frac{1}{\sqrt{1-\frac{v^2}{c^2}}} \tag{5.1}$$

relative to its rest mass, which would result in an infinite mass in the case of $v = c$.

If a mass is accelerated with a constant force, its velocity increases according to

$$F = m \cdot a \tag{5.2}$$

$$F = m \cdot \frac{\Delta v}{\Delta t} \tag{5.3}$$

$$\Delta v = \frac{F \cdot \Delta t}{m} \tag{5.4}$$

If F, Δt, and m are constant, then Δv is also constant.

However, if the mass increases under constant applied force according to

$$m_{accelerated} = \frac{m_{Quiet}}{\sqrt{1-\frac{v^2}{c^2}}} \tag{5.5}$$

J. von Stackelberg, *The Mass of a Photon*, https://doi.org/10.1007/978-3-658-49408-7_5

then the acceleration decreases according to

$$a_{accelerated} = a_{Quiet} \cdot \sqrt{1 - \frac{v^2}{c^2}} \tag{5.6}$$

i.e., the closer the velocity v approaches the limiting velocity c, the lower the acceleration becomes.

A similar effect occurs when the applied force decreases with increasing velocity according to

$$F_{accelerated} = F_{Quiet} \cdot \sqrt{1 - \frac{v^2}{c^2}} \tag{5.7}$$

This effect arises when the element causing the applied force, such as a force field, moves close to or at the limiting velocity.

An example: If you push a car with another car, you can at most reach the speed that the pushing car is capable of achieving.

If these acting fields responsible for accelerating and detecting the mass all move at the speed of light, then it is not possible to accelerate a mass beyond this limiting velocity, nor is it possible to determine to what extent the mass has changed.

If one wanted to accelerate mass to a velocity greater than the speed of light, a mechanism would have to be found whose speed is independent of the propagation of waves. An analogous function can be seen in aircraft development: propeller-driven aircraft were unable to break the sound barrier because propellers pull the aircraft through the air by utilizing the flow and pressure conditions in the air, similar to sound. Jet-powered aircraft, on the other hand, are based on a different principle, namely the expansion of gas volume due to combustion, and are thus apparently capable of achieving multiple times the speed of sound. Investigations in this regard, concerning the acceleration of masses, should address the formation of the Bose–Einstein condensate [3].

References

1. Jay O'Rear; Physik; Carl Hanser Verlag München; 1989
2. Albert Einstein; Relativity – The Special & the General Theory; Martino Publishing Mansfield Centre; 2010
3. Wikipedia, Bose-Einstein-Kondensat, 1. 3. 2021

6 Mass

Gravitational fields, like electromagnetic fields, propagate at the speed of light. In Maxwell's equations, the material equations (see Sect. 4.1.4) contain the two constants μ_0 for permeability and ε_0 for permittivity. Taken together, these constants yield the speed of light in vacuum. Considered individually, and with appropriate interpretation, the constants describe the propagation of the magnetic and electric fields, respectively. However, the original derivation of these constants did not take this relationship into account. With a corresponding reassessment of quantitative mass relative to the currently valid mass value, both constants could potentially be omitted entirely.

Suppose a filamentary electrical conductor is traversed by a time-varying electric current I [1]:

$$I(t) = I_0 \cdot \sin \omega \cdot t \tag{6.1}$$

with

$$S = \frac{d}{dA} I = \frac{d}{dt} D \tag{6.2}$$

S = Current density in $\frac{A}{m^2}$
D = Electric displacement density in $\frac{As}{m^2}$

A circular field with magnetic field strength then forms around the conductor

$$H(t) = \frac{I_0 \cdot \cos \omega \cdot t}{l_m} \tag{6.3}$$

J. von Stackelberg, *The Mass of a Photon*, https://doi.org/10.1007/978-3-658-49408-7_6

with

$l_m =$ the length of the magnetic field line in m

This means that the field strength varies linearly and inversely with the distance from the conductor.

The magnetic field strength generates a magnetic flux density

$$B(t) = \mu_0 \cdot H(t) \tag{6.4}$$

$$B(t) = \mu_0 \cdot \frac{I_0 \cdot \cos \omega \cdot t}{l_m} \tag{6.5}$$

The permeability constant μ_0 also contains a temporal dependence, as indicated by its unit:

$$\frac{s}{m} \cdot \frac{V}{A}, \tag{6.6}$$

i.e., the magnetic flux arises from the magnetic field strength with a temporal delay, which may be

$$4 \cdot \pi \cdot 10^{-7} \frac{s}{m} \cdot \frac{V}{A} \tag{6.7}$$

.

Closed electric field lines form around the circular lines of the time-varying magnetic flux:

$$E = -\frac{d}{dt} B \tag{6.8}$$

$$E = -\frac{d}{dt}\left(\mu_0 \cdot \frac{I_0 \cdot \cos \omega \cdot t}{l_m}\right) \tag{6.9}$$

$$E = \mu_0 \cdot \omega \cdot \frac{I_0 \cdot \sin \omega \cdot t}{l_m} \tag{6.10}$$

The direction of the electric field can be seen in Fig. 6.1.

The electric displacement arises from the electric field strength:

$$D = \varepsilon_0 \cdot E \tag{6.11}$$

$$D = \varepsilon_0 \cdot \mu_0 \cdot \omega \cdot \frac{I_0 \cdot \sin \omega \cdot t}{l_m} \tag{6.12}$$

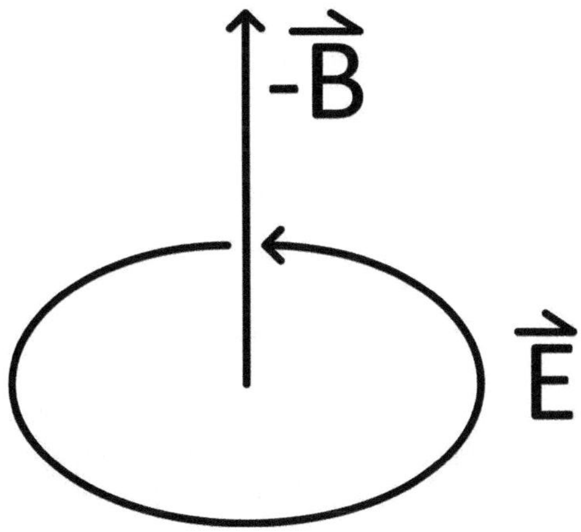

Fig. 6.1 Determination of the direction of action of the electric field from the direction of the magnetic flux density

The permittivity constant ε_0 also exhibits temporal dependence:

$$\varepsilon_0 = 8{,}85 \cdot 10^{-12} \frac{s}{m} \cdot \frac{A}{V} \tag{6.13}$$

In principle, spatial field propagation occurs according to

$$\frac{1}{r^2}, \tag{6.14}$$

since the surface area of a sphere is calculated as follows:

$$O = 4 \cdot \pi \cdot r^2 \tag{6.15}$$

In this context, the factor $4 \cdot \pi$, which appears in the expression for the electric field strength around a charged sphere, is also noteworthy:

$$E = \frac{q}{\varepsilon_0} \cdot \frac{1}{4 \cdot \pi \cdot r^2} \tag{6.16}$$

I wonder whether the $4 \cdot \pi$ that appears in the permeability constant is also related to the way a field develops as a function of distance from its source.

The following questions and assumptions arise from this consideration:

F1: If a certain current I with the property

$$I(t) = I_0 \cdot \sin \omega \cdot t \tag{6.17}$$

flows through a piece of conductor for an infinite period of time, and this infinite duration of current flow pumps an infinite amount of electromagnetic energy in the form of electromagnetic waves into infinite space, does the amplitude of the electromagnetic field at a given distance r from the conductor change over time, or does the amplitude remain constant?

A1: The amplitude remains constant, because the amount of energy flowing to a given point continues to flow onward at the same rate; however, the distance at which a field can still be detected increases over time, since the field propagates at the speed of light.

F2: What is the propagation speed of an electric and a magnetic field?

A2: The propagation speed is equal to the (vacuum) speed of light.

F3: What if the speed at which photons travel is different from the speed at which planar or spatial electromagnetic waves propagate, and this in turn differs from the propagation speed of an electric or magnetic field?

A3: That is not the case.

F4: Why does the permeability constant include the factor $4 \cdot \pi$?

A4: Because the permeability constant was introduced without reference to physical conditions.

F5: What would happen to the permeability constant and the permittivity constant if the unit of length "one meter" were replaced by another unit of length?

A5: Their values would change.

F6: Is the difference in phase shift between the electric and magnetic fields in the near and far field [2] due to measurement findings, or does the phase shift arise purely mathematically, for example due to inadequate models?

A6: It arises purely mathematically (see also Sect. 4.1.5).

Justifications for the above assumptions:

The current I flowing through the conductor (or the electric displacement density D in the medium) IS EQUAL to the magnetic field strength H around the conductor or in the medium. Therefore, there is no temporal or spatial relationship between the two. Conversely, the electric field strength E is, of course, EQUAL to the magnetic flux density B, with respect to their temporal gradients, since a magnetic flux that is temporally constant does not generate an electric field. The temporal relationship arises only from the electric field strength (or electric voltage as the line integral of the electric field strength) with respect to the current flow or the electric displacement density.

At the same time, I assume that the propagation speed of a magnetic field around an electric conductor, the propagation speed of electromagnetic waves, and the speed of light are identical. With reference to the vacuum in our system of units and magnitudes, this is approximately $3 \cdot 10^8$ m/s. It should be noted that if a different unit of length is used, this value will also take on a different magnitude.

In addition to voltage and current dependencies, the permittivity constant and the permeability constant also include dependencies on time and length, as can be easily seen from the units As/Vm and Vs/Am, respectively.

Due to the development of physics and electrical engineering, both constants have been assigned more or less arbitrary values, which themselves result from other quantities that were also defined more or less arbitrarily. At the same time,

$$c = \frac{1}{\sqrt{\mu_0 \cdot \varepsilon_0}} \tag{6.18}$$

applies, with the current values

$$\mu_0 = 4 \cdot \pi \cdot 10^{-7} \frac{V}{A} \cdot \frac{s}{m} \tag{6.19}$$

$$\varepsilon_0 = 8{,}85 \cdot 10^{-12} \frac{A}{V} \cdot \frac{s}{m} \tag{6.20}$$

If I now unify the two constants in such a way that

$$\mu_{ovst} = \varepsilon_{ovst} = \frac{1}{c} = \frac{1}{3 \cdot 10^8} \left(\frac{V \cdot s}{A \cdot m} \quad \text{bzw.} \quad \frac{A \cdot s}{V \cdot m} \right), \tag{6.21}$$

how would the rest of electrical engineering and physics have to be rewritten in order to yield a consistent system again? The major advantage of this approach would be that two "fundamental constants" could be eliminated entirely, thereby simplifying the overall system.

The voltage in V(olts) is a derived unit, which is calculated from

$$P_a = P_{mech} \tag{6.22}$$

$$U \cdot I = \frac{m \cdot s^2}{t^3} \tag{6.23}$$

$$U = \frac{m \cdot s^2}{t^3 \cdot I} \tag{6.24}$$

The mass appears one-dimensionally in the equation. If the mass were to be assigned the correction factor, this would mean that the previously valid reference for mass—one kilogram as one cubic decimeter of water at 4 °C—would lose its basis.

$$\mu_{0vSt} = \frac{1}{c} = \frac{1}{3 \cdot 10^8} \frac{V \cdot s}{A \cdot m} \tag{6.25}$$

$$\mu_0 = 4 \cdot \pi \cdot 10^{-7} = a_\mu \cdot \mu_{0vSt} = a_\mu \cdot \frac{1}{3 \cdot 10^8} \frac{V \cdot s}{A \cdot m} \tag{6.26}$$

$$a_\mu = \frac{4 \cdot \pi \cdot 10^{-7}}{\frac{1}{3 \cdot 10^8}} \tag{6.27}$$

$$a_\mu = 120 \cdot \pi \tag{6.28}$$

$$\mu_{0vst} = \frac{1}{c} = \frac{\mu_0}{a_\mu} = \frac{\mu_0}{120 \cdot \pi} \tag{6.29}$$

$$\varepsilon_{0vSt} = \frac{1}{c} = \frac{1}{3 \cdot 10^8} \frac{A \cdot s}{V \cdot m} \tag{6.30}$$

$$\varepsilon_0 = 8{,}85 \cdot 10^{-12} = a_\varepsilon \cdot \varepsilon_{0vSt} = a_\varepsilon \cdot \frac{1}{3 \cdot 10^8} \frac{A \cdot s}{V \cdot m} \tag{6.31}$$

$$a_\varepsilon = \frac{8{,}85 \cdot 10^{-12}}{\frac{1}{3 \cdot 10^8}} = \frac{\frac{1}{9 \cdot 10^{16} \cdot 4 \cdot \pi \cdot 10^{-7}}}{\frac{1}{3 \cdot 10^8}} \tag{6.32}$$

$$a_\varepsilon = \frac{1}{3 \cdot 10^8 \cdot 4 \cdot \pi \cdot 10^{-7}} \tag{6.33}$$

$$a_\varepsilon = \frac{1}{120 \cdot \pi} \tag{6.34}$$

$$\varepsilon_{0vSt} = \frac{1}{c} = \varepsilon_0 \cdot 120 \cdot \pi \tag{6.35}$$

This change in the two constants has consequences in many areas of electrical engineering:

If the electric current I in A(mperes) is left unchanged, since it is a base unit and is defined by the force (which is also mass times acceleration on the mechanical side) acting on the two conductors through which the current flows, then the magnetic field strength H also remains unchanged. The first change occurs in the representation of the magnetic flux, which is reduced in value by a factor of $120 \cdot \pi$:

$$B_{vst}(t) = \mu_{ovSt} \cdot \frac{I_0 \cdot \cos \omega \cdot t}{l_m} \tag{6.36}$$

$$B_{vst}(t) = \frac{\mu_0}{120 \cdot \pi} \cdot \frac{I_0 \cdot \cos \omega \cdot t}{I_m} \tag{6.37}$$

As a result, the electric field strength is also reduced by a factor of $120 \cdot \pi$:

$$E_{vSt} = \mu_{0vSt} \cdot \omega \cdot \frac{I_0 \cdot \sin \omega \cdot t}{l_m} \tag{6.38}$$

$$E_{vSt} = \frac{\mu_0}{120 \cdot \pi} \cdot \omega \cdot \frac{I_0 \cdot \sin \omega \cdot t}{l_m} \tag{6.39}$$

This has a direct effect on the electric voltage, which is likewise reduced by a factor of $120 \cdot \pi$:

$$U_{vst} = E_{vst} \cdot s \tag{6.40}$$

This presents a problem, since the transition from mechanical to electrical power is no longer maintained. One possible solution would be to define electrical power as follows:

$$P_{el} = U_{vst} \cdot I \cdot 120 \cdot \pi \tag{6.41}$$

A better alternative is to introduce a new reference value for mass.

With

$$m_{vSt} = \frac{m}{120 \cdot \pi} \tag{6.42}$$

it is possible to simultaneously reconcile the transitions between mechanics and electrical engineering that arise due to the Coulomb force and the Lorentz force.

From the reduced electric field strength, the electric displacement density is obtained, which remains unchanged when using the corrected value for the permittivity constant:

$$D_{vSt} = \varepsilon_{0vSt} \cdot \mu_{0vSt} \cdot \omega \cdot \frac{I_0 \cdot \sin \omega \cdot t}{l_m} \tag{6.43}$$

$$D_{vSt} = \varepsilon_0 \cdot 120 \cdot \pi \cdot \frac{\mu_0}{120 \cdot \pi} \cdot \omega \cdot \frac{I_0 \cdot \sin \omega \cdot t}{l_m} \tag{6.44}$$

$$D_{vSt} = D = \varepsilon_0 \cdot \mu_0 \cdot \omega \cdot \frac{I_0 \cdot \sin \omega \cdot t}{l_m} \tag{6.45}$$

Furthermore, this approach affects the values of the following physical quantities:

- Coulomb force, which is reduced by a factor of 120 · π:

$$F_C = \frac{q_1 \cdot q_2}{\varepsilon_0} \cdot \frac{1}{4 \cdot \pi \cdot r^2} \tag{6.46}$$

$$F_{CvSt} = \frac{q_1 \cdot q_2}{\varepsilon_{0ust}} \cdot \frac{1}{4 \cdot \pi \cdot r^2} \tag{6.47}$$

$$F_{CvSt} = \frac{q_1 \cdot q_2}{\varepsilon_0 \cdot 120 \cdot \pi} \cdot \frac{1}{4 \cdot \pi \cdot r^2} \tag{6.48}$$

- Capacitance of a capacitor, which is increased by a factor of 120 · π:

$$C_{vSt} = \frac{\varepsilon_{0vSt} \cdot \varepsilon_r \cdot A}{d} \tag{6.49}$$

$$C_{vSt} = \frac{\varepsilon_0 \cdot 120 \cdot \pi \cdot \varepsilon_r \cdot A}{d} \tag{6.50}$$

As a result, the capacitive reactance also changes:

$$X_{CYSt} = \frac{1}{2 \cdot \pi \cdot f \cdot C_{vSt}} \tag{6.51}$$

$$X_{CvSt} = \frac{1}{2 \cdot \pi \cdot f \cdot C \cdot 120 \cdot \pi} \tag{6.52}$$

However, it is compensated again by the reduced voltage:

$$X_{CvSt} = \frac{U_{Cvst}}{I_C} \tag{6.53}$$

$$\frac{X_C}{120 \cdot \pi} = \frac{\frac{U_C}{120 \cdot \pi}}{I_C} \tag{6.54}$$

$$\frac{X_C}{120 \cdot \pi} = \frac{U_C}{I_C \cdot 120 \cdot \pi} \tag{6.55}$$

- Magnetic force between two conductors based on the Lorentz force:

$$\vec{F_L} = q \cdot \vec{v} \times \vec{B} \tag{6.56}$$

For appropriately perpendicular conditions, the vectors become scalars and the cross product becomes a multiplication:

$$F_L = q \cdot v \cdot B \tag{6.57}$$

$$F_{Lvst} = q \cdot v \cdot B_{vSt} \tag{6.58}$$

$$F_{LvSt} = q \cdot v \cdot \frac{\mu_0}{120 \cdot \pi} \cdot H \tag{6.59}$$

The Lorentz force on a charge is also reduced by a factor of $120 \cdot \pi$, thus matching the Coulomb force.

- Inductance of a coil, which is reduced by a factor of $120 \cdot \pi$, and thus the inductive reactance, which, however, is again compensated by the reduced voltage,
- Reduced ohmic resistance due to the reduced voltage,
- The resonance frequencies of LC combinations remain unchanged in value, since the changes in L and C cancel each other out,
- The cutoff frequencies of RC and RL combinations remain unchanged, because the changes in R and C or L compensate each other,

New approach regarding possible changes to the fundamental quantities of physics.

The seven independent base quantities of physics

- Length s in meter m
- Mass m in kilogram kg
- Time t in second s
- Temperature T in kelvin K
- Electric current I in ampere A
- Amount of substance in mole mol
- Luminous intensity in candela cd

originated from, in part, much older quantities that were unified, and in their original definitions have an obvious connection to everyday dimensions.

The unit meter (m) was defined as one ten-millionth of the distance from the North Pole to the Equator along the meridian passing through Paris [3; 4]. This length was measured in the 18th century by two French scientists. The modern definition, that a meter is the length that light travels in vacuum in 1/299,792,458 s, is therefore a derived definition, not least because another base quantity, namely time, must also be included.

The kilogram as a unit of mass was derived, after the definition of the meter as the unit of length, as the mass of one cubic decimeter of water at its maximum density (4 °C) [5]. Currently, a redefinition of the kilogram is being developed, based on a specific number of atoms in a monocrystal, where, again, the macroscopic accuracy of mass determination depends on the accuracy of length measurement (measurement of the crystal's diameter).

The second as a unit of time was defined as the sixtieth part of the sixtieth part of the twenty-fourth part of a mean Earth rotation, i.e., the duodecimal system plays a central role here [6]. The modern definition of time measurement refers to the oscillation of a cesium atom at a specific energy transition.

The kelvin as a unit of temperature was originally derived from the linear scaling from 0 to 100 of the two phase transitions of water from solid to liquid (0 °C) and from liquid to gaseous (100 °C) at standard atmospheric pressure (air pressure = 1013 mbar). The zero point of the kelvin scale is at −273.15 °C.

The electric current of one ampere was originally defined as the amount of current that deposits 1.118 mg of silver from a silver nitrate solution by electrolysis in one second. The modern definition states that a current of one ampere, when flowing through two infinitely long conductors placed one meter apart, produces a force of 2×10^{-7} newton per meter of conductor length between these conductors.

The amount of substance in mole corresponds to the number of elementary entities contained in 12 g of the carbon nuclide ^{12}C in the unbound state.

The luminous intensity in candela refers to the luminous flux passing through this area per steradian.

Derived units: Originally, electric voltage was derived from the voltage generated in a so-called Weston standard cell. The Weston standard cell is a galvanic (primary) cell that produces a specific voltage through a chemical process between two conductor materials connected by an electrolyte.

With regard to the problem of unifying permeability, permittivity, and the speed of light, the following quantities are considered:

- Electric voltage,
- Electric current,
- Mass,
- Length,
- Time,
- Power

Power is included because it links the units of electrical engineering with those of mechanics. Through this connection, the magnitude of electric voltage is defined in the modern world.

Ultimately, the quantities listed above are either dependent on length in various dimensions or independent:

- Electric voltage: Independent
- Electric current: Dependent in the third dimension (mass dependence)
- Mass: Dependent in the third dimension
- Length: Base unit

- Time: Independent
- Power: Dependent in the fifth dimension (mass and length dependence)

Qualitatively, the following dependencies can be identified:

- The value of the speed of light depends reciprocally and linearly on the magnitude of length, i.e., increasing the reference length by a factor decreases the value of the speed of light by the same factor.
- The value of mass depends reciprocally in the third dimension on the magnitude of length, i.e., increasing the reference length by a factor decreases the value of mass by that factor cubed.
- The value of current depends reciprocally and linearly on the magnitude of mass, i.e., increasing the reference mass by a factor decreases the value of current by that factor. Thus, the value of current in the third dimension depends on the magnitude of length, i.e., increasing the reference length by a factor increases the value of current by that factor cubed.
- The value of mechanical power depends reciprocally in the fifth dimension on the magnitude of length, i.e., increasing the reference length by a factor decreases the value of power by that factor to the fifth power.

If one examines the interfaces between mechanical and electrical quantities, especially with regard to the question of whether and to what extent the value change by the factor $120 \cdot \pi$ plays a role or is consistent, it becomes apparent that the focus is on forces and on power or work. In view of these changes, it may be reasonable to reconsider the definition of mass:

$$m_{vSt} = \frac{m}{120 \cdot \pi} \tag{6.60}$$

References

1. Ingo Wolff; Grundlagen und Anwendungen der Maxwellschen Theorie I – Ein Repetitorium; VDI Verlag; 1996
2. Alfred Fettweis; Electrical communications, fluid dynamics, and some fundamental issues in physics; Ferdinand Schöningh; 2010
3. Wikipedia, Meter, 1. 3. 2021
4. Daniel Kehl; Die Vermessung der Welt; Rororo Verlag, 2008
5. Wikipedia, Kilogramm, 1. 3. 2021
6. Wikipedia, Sekunde, 1. 3. 2021

7 Fourier Series

Due to my professional responsibilities, I had to deal with the generation of harmonic overtones in chopped sinusoidal voltages and currents. I attempted to theoretically resolve the question of the number of harmonics involved—and failed, as will be described in detail.

With the aid of Fourier series, any periodic function can be represented as a sum of harmonic sinusoidal oscillations; at least, this is the prevailing view in the literature. The formalism appears straightforward [1]:

$$f(t) = \frac{a_0}{2} + \sum_{k=1}^{\infty} (a_k \cdot \cos(k \cdot t) + b_k \cdot \sin(k \cdot t)) \tag{7.1}$$

with

$$a_k = \frac{1}{\pi} \int_{-\pi}^{\pi} f(t) \cdot \cos(k \cdot t) dt \quad \text{for} \quad k \geq 0 \tag{7.2}$$

$$b_k = \frac{1}{\pi} \int_{-\pi}^{\pi} f(t) \cdot \sin(k \cdot t) dt \quad \text{for} \quad k \geq 1 \tag{7.3}$$

The most frequently used example in the literature to demonstrate the application of Fourier series is the rectangular function, whose period segment around the origin can be described as follows:

$$f(t) = \begin{cases} -1 & \text{for} \quad t \in [-\pi; 0[\\ +1 & \text{for} \quad t \in [0; \pi[\end{cases} \tag{7.4}$$

J. von Stackelberg, *The Mass of a Photon*, https://doi.org/10.1007/978-3-658-49408-7_7

If this function is substituted into the expressions for a_k and b_k, the following results are obtained:

$$a_k = \frac{1}{\pi} \cdot \int_{-\pi}^{\pi} f(t) \cdot \cos(k \cdot t)dt \text{ for } k \geq 0 \tag{7.5}$$

$$a_k = \frac{1}{\pi} \cdot \left(\int_{-\pi}^{0} -1 \cdot \cos(k \cdot t)dt + \int_{0}^{\pi} 1 \cdot \cos(k \cdot t)dt \right) \tag{7.6}$$

$$a_k = \frac{1}{\pi} \cdot \left(-\int_{-\pi}^{0} \cos(k \cdot t)dt + \int_{0}^{\pi} \cos(k \cdot t)dt \right) \tag{7.7}$$

$$a_k = \frac{1}{\pi} \cdot \left(-\left[\frac{1}{k} \cdot \sin(k \cdot t) \right]_{-\pi}^{0} + \left[\frac{1}{k} \cdot \sin(k \cdot t) \right]_{0}^{\pi} \right) \tag{7.8}$$

$$a_k = \frac{1}{\pi} \cdot \frac{1}{k} \cdot (-[\sin(k \cdot 0) - \sin(k \cdot (-\pi))] + [\sin(k \cdot \pi) - \sin(k \cdot 0)]) \tag{7.9}$$

$$a_k = 0 \tag{7.10}$$

$$b_k = \frac{1}{\pi} \cdot \int_{-\pi}^{\pi} f(t) \cdot \sin(k \cdot t)dt \quad \text{for} \quad k \geq 1 \tag{7.11}$$

$$b_k = \frac{1}{\pi} \cdot \left(\int_{-\pi}^{0} -1 \cdot \sin(k \cdot t)dt + \int_{0}^{\pi} 1 \cdot \sin(k \cdot t)dt \right) \tag{7.12}$$

$$b_k = \frac{1}{\pi} \cdot \left(-\left[-\frac{1}{k} \cdot \cos(k \cdot t) \right]_{-\pi}^{0} + \left[-\frac{1}{k} \cdot \cos(k \cdot t) \right]_{0}^{\pi} \right) \tag{7.13}$$

$$b_k = -\frac{1}{\pi} \cdot \frac{1}{k} \cdot (-[\cos(k \cdot 0) - \cos(k \cdot (-\pi))] + [\cos(k \cdot \pi) - \cos(k \cdot 0)]) \tag{7.14}$$

$$b_k = -\frac{1}{\pi} \cdot \frac{1}{k} \cdot (-\cos(k \cdot 0) - \cos(k \cdot 0) + \cos(k \cdot (-\pi)) + \cos(k \cdot \pi)) \tag{7.15}$$

with

$$\cos(k \cdot (-\pi)) = \cos(k \cdot \pi) \tag{7.16}$$

applies

$$b_k = \frac{2}{k \cdot \pi} \cdot (\cos(k \cdot 0) - \cos(k \cdot \pi)) \tag{7.17}$$

for even k

$$b_k = 0 \tag{7.18}$$

for odd k

$$b_k = \frac{4}{k \cdot \pi} \tag{7.19}$$

Substituting into

$$f(t) = \frac{a_0}{2} + \sum_{k=1}^{\infty} (a_k \cdot \cos(k \cdot t) + b_k \cdot \sin(k \cdot t)) \tag{7.20}$$

yields

$$f(t) = \frac{4}{\pi} \cdot \sum_{k=1}^{\infty} \frac{1}{(2 \cdot k - 1)} \cdot \sin((2 \cdot k - 1) \cdot t) \tag{7.21}$$

If you program a corresponding table in Excel, you obtain a graph as shown in Fig. 7.1:

Next, a function describing a pulse-width modulated signal will be examined.

The function for a rectangular waveform with variable duty cycle can be described as follows:

$$f(t) = \frac{a_0}{2} + \sum_{k=1}^{\infty} (a_k \cdot \cos(k \cdot t) + b_k \cdot \sin(k \cdot t)) \tag{7.22}$$

$$a_k = \frac{2}{T} \int_c^{c+T} f(t) \cdot \cos(k \cdot t) dt \tag{7.23}$$

$$b_k = \frac{2}{T} \int_c^{c+T} f(t) \cdot \sin(k \cdot t) dt \tag{7.24}$$

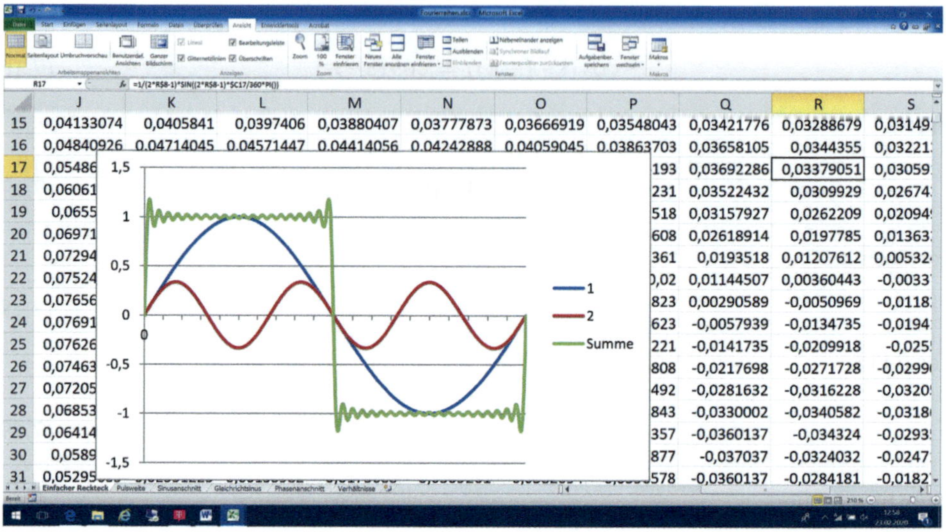

Fig. 7.1 Graph of the fundamental and the third harmonic together with the sum curve up to the 20th harmonic according to the Fourier series for a square wave; this graph confirms the information found in textbooks

$$f(t) = \begin{cases} a & \text{for} \quad t \in [0; z] \quad \text{with} \quad z \in [0 \cdot T; 1 \cdot T] \\ 0 & \text{for} \quad t \in]z; T] \quad \text{with} \quad z \in [0 \cdot T; 1 \cdot T] \end{cases} \tag{7.25}$$

$$a_k = \frac{2}{T} \int_0^z a \cdot \cos(k \cdot t) dt \tag{7.26}$$

$$a_k = \frac{2 \cdot a}{T} \cdot \int_0^z \cos(k \cdot t) dt \tag{7.27}$$

$$a_k = \frac{2 \cdot a}{k \cdot T} \cdot [\sin(k \cdot t)]_0^z \tag{7.28}$$

$$a_k = \frac{2 \cdot a}{k \cdot T} \cdot \sin(k \cdot z) \tag{7.29}$$

for k = 0:

$$a_0 = \lim_{k \to 0} \frac{2 \cdot a \cdot \sin(k \cdot z)}{k \cdot T} \tag{7.30}$$

Applying l'Hospital's rule:

$$a_0 = \lim_{k \to 0} \frac{2 \cdot a \cdot z \cdot \cos(k \cdot z)}{T} \tag{7.31}$$

$$a_0 = \frac{2 \cdot a \cdot z}{T} \tag{7.32}$$

$$b_k = \frac{2}{T} \cdot \int_0^z a \cdot \sin(k \cdot t)dt \tag{7.33}$$

$$b_k = -\frac{2 \cdot a}{k \cdot T} \cdot [\cos(k \cdot t)]_0^z \tag{7.34}$$

$$b_k = -\frac{2 \cdot a}{k \cdot T} \cdot (\cos(k \cdot z) - 1) \tag{7.35}$$

$$b_k = \frac{2 \cdot a}{k \cdot T} \cdot (1 - \cos(k \cdot z)) \tag{7.36}$$

$$f(t) = \frac{a \cdot z}{T} + \frac{2 \cdot a}{T} \cdot \sum_{k=1}^{\infty} \frac{1}{k} \cdot (\sin(k \cdot z) \cdot \cos(k \cdot t) - \cos(k \cdot z) \cdot \sin(k \cdot t) + \sin(k \cdot t)) \tag{7.37}$$

$$f(t) = \frac{a \cdot z}{T} + \frac{2 \cdot a}{T} \cdot \sum_{k=1}^{\infty} \frac{1}{k} \cdot (\sin(k \cdot z - k \cdot t) + \sin(k \cdot t)) \tag{7.38}$$

$$f(t) = \frac{a \cdot z}{T} + \frac{2 \cdot a}{T} \cdot \sum_{k=1}^{\infty} \frac{1}{k} \cdot (\sin(k \cdot (z - t)) + \sin(k \cdot t)) \tag{7.39}$$

For $z = 0.1$ T, an Excel plot is obtained as shown in Fig. 7.2.

For $z = 0.5$ T, an Excel plot is obtained as shown in Fig. 7.3.

For $z = 0.9$ T, an Excel plot is obtained as shown in Fig. 7.4.

A rather unusual, but nonetheless common signal in practice is the sinusoidal signal phase-controlled by a thyristor set (Fig. 7.5).

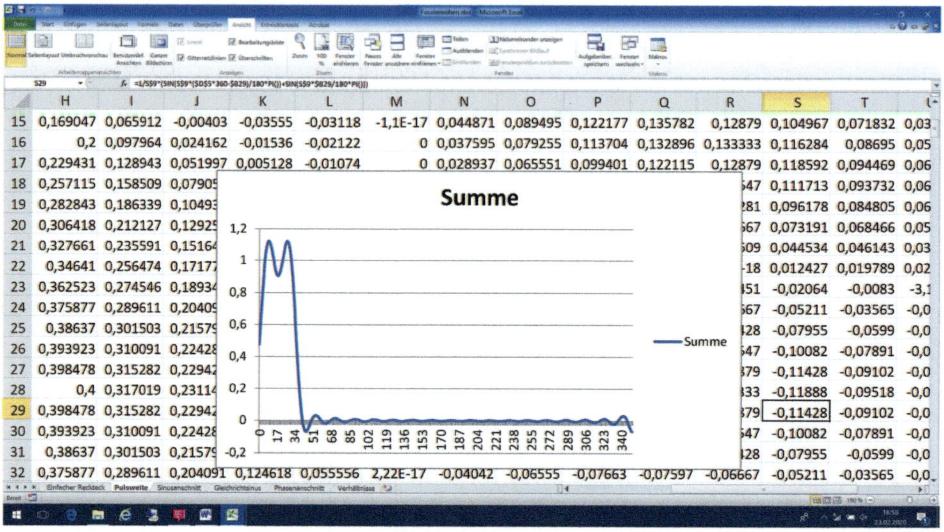

Fig. 7.2 Plot of the sum of the Fourier series up to the 20th harmonic of a pulse-width modulated signal with a pulse duration of 10 %

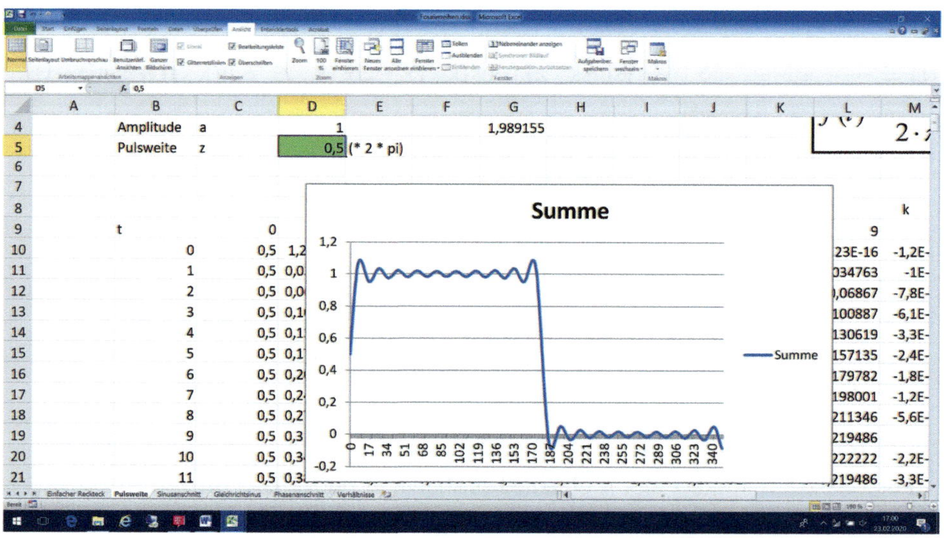

Fig. 7.3 Plot of the sum of the Fourier series up to the 20th harmonic of a pulse-width modulated signal with a pulse duration of 50 %

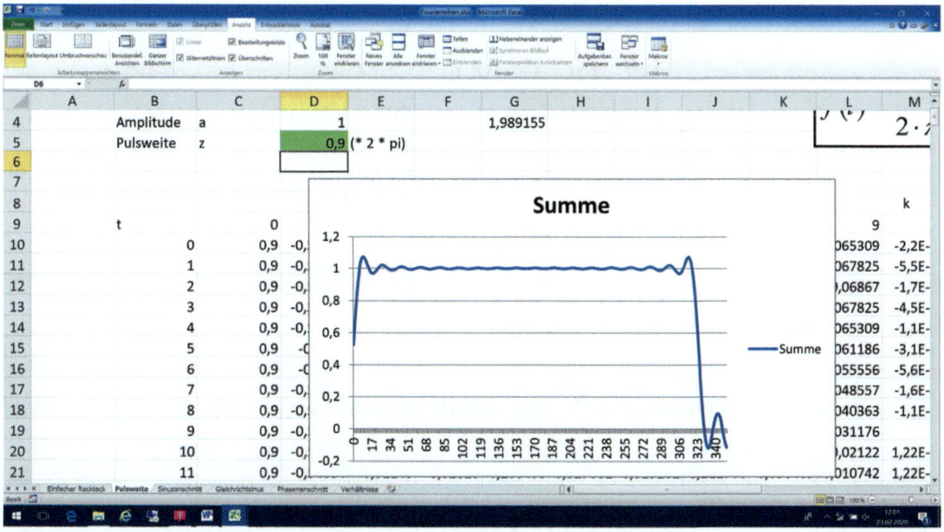

Fig. 7.4 Plot of the sum of the Fourier series up to the 20th harmonic of a pulse-width modulated signal with a pulse duration of 90 %

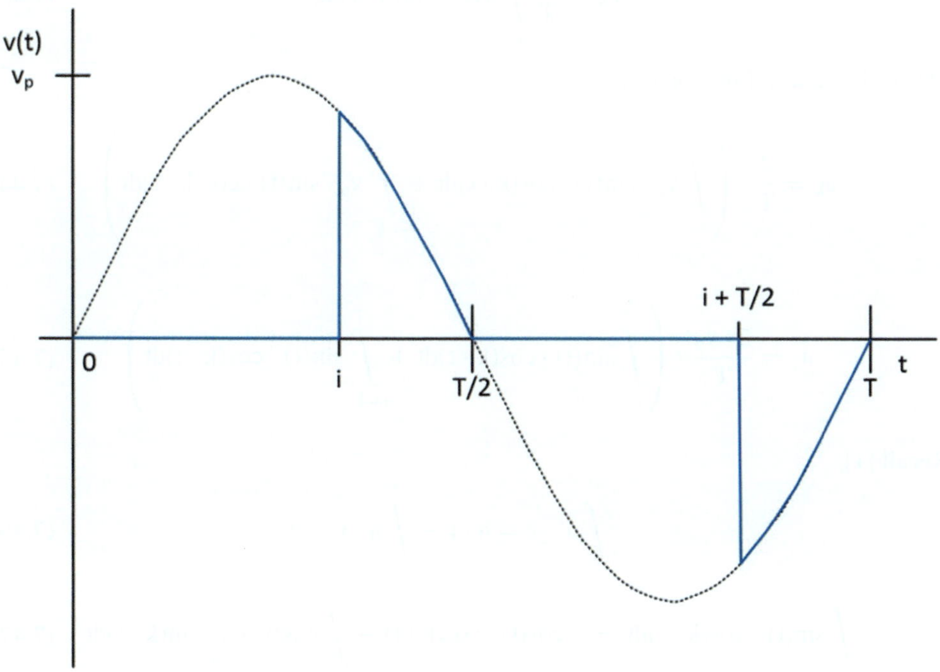

Fig. 7.5 Plot of a sinusoidal signal phase-controlled on both half-waves at the same angle by a thyristor set

The function is described as follows:

$$f(t) = \begin{cases} 0 \quad \text{for} \quad t \in [0; i[\\ v_p \cdot \sin(2 \cdot \pi \cdot f \cdot t) \quad \text{for} \quad t \in \left[i; \frac{T}{2}\right[\\ 0 \quad \text{for} \quad t \in \left[\frac{T}{2}; i + \frac{T}{2}\right[\\ v_p \cdot \sin(2 \cdot \pi \cdot f \cdot t) \quad \text{for} \quad t \in \left[i + \frac{T}{2}; T\right[\end{cases} \tag{7.40}$$
$$\text{with} \quad i \in \left[0\% \cdot \frac{T}{2}; 100\% \cdot \frac{T}{2}\right[$$

The Fourier series is generated as follows:

$$f(t) = \frac{a_0}{2} + \sum_{k=1}^{\infty} (a_k \cdot \cos(k \cdot t) + b_k \cdot \sin(k \cdot t)) \tag{7.41}$$

$$a_k = \frac{2}{T} \int_c^{c+T} f(t) \cdot \cos(k \cdot t) dt \tag{7.42}$$

$$b_k = \frac{2}{T} \int_c^{c+T} f(t) \cdot \sin(k \cdot t) dt \tag{7.43}$$

Determine a_k as a function of i:

$$a_k = \frac{2}{T} \cdot \left(\int_i^{\frac{T}{2}} v_p \cdot \sin(t) \cdot \cos(k \cdot t) dt + \int_{i+\frac{T}{2}}^{T} v_p \cdot \sin(t) \cdot \cos(k \cdot t) dt \right) \tag{7.44}$$

$$a_k = \frac{2 \cdot v_p}{T} \cdot \left(\int_i^{\frac{T}{2}} \sin(t) \cdot \cos(k \cdot t) dt + \int_{i+\frac{T}{2}}^{T} \sin(t) \cdot \cos(k \cdot t) dt \right) \tag{7.45}$$

Recall [1]:

$$\int u' \cdot v = u \cdot v - \int u \cdot v' \tag{7.46}$$

$$\int \sin(t) \cdot \cos(k \cdot t) dt = -\cos(t) \cdot \cos(k \cdot t) - \int \cos(t) \cdot k \cdot \sin(k \cdot t) dt \tag{7.47}$$

$$\int \sin(t) \cdot \cos(k \cdot t) dt = -\cos(t) \cdot \cos(k \cdot t) - k \cdot \int \cos(t) \cdot \sin(k \cdot t) dt \tag{7.48}$$

$$\int \sin(t)\cdot\cos(k\cdot t)dt = -\cos(t)\cdot\cos(k\cdot t) - k\cdot\left[\sin(t)\cdot\sin(k\cdot t) - \int \sin(t)\cdot k\cdot\cos(k\cdot t)dt\right] \tag{7.49}$$

$$\int \sin(t)\cdot\cos(k\cdot t)dt = -\cos(t)\cdot\cos(k\cdot t) - k\cdot\left[\sin(t)\cdot\sin(k\cdot t) - k\cdot\int \sin(t)\cdot\cos(k\cdot t)dt\right] \tag{7.50}$$

$$\int \sin(t)\cdot\cos(k\cdot t)dt = -\cos(t)\cdot\cos(k\cdot t) - k\cdot\sin(t)\cdot\sin(k\cdot t) + k^2\cdot\int \sin(t)\cdot\cos(k\cdot t)dt \tag{7.51}$$

$$\int \sin(t)\cdot\cos(k\cdot t)dt - k^2\cdot\int \sin(t)\cdot\cos(k\cdot t)dt = -\cos(t)\cdot\cos(k\cdot t) - k\cdot\sin(t)\cdot\sin(k\cdot t) \tag{7.52}$$

$$\left(1-k^2\right)\cdot\int \sin(t)\cdot\cos(k\cdot t)dt = -\cos(t)\cdot\cos(k\cdot t) - k\cdot\sin(t)\cdot\sin(k\cdot t) \tag{7.53}$$

$$\int \sin(t)\cdot\cos(k\cdot t)dt = \frac{1}{k^2-1}\cdot(\cos(t)\cdot\cos(k\cdot t) + k\cdot\sin(t)\cdot\sin(k\cdot t)) \tag{7.54}$$

$$a_k = \frac{1}{k^2-1}\cdot\frac{2\cdot v_p}{T}\cdot\begin{pmatrix}[\cos(t)\cdot\cos(k\cdot t)+k\cdot\sin(t)\cdot\sin(k\cdot t)]_i^{\frac{T}{2}}+\\ +[\cos(t)\cdot\cos(k\cdot t)+k\cdot\sin(t)\cdot\sin(k\cdot t)]_{i+\frac{T}{2}}^{T}\end{pmatrix} \tag{7.55}$$

$$a_k = \frac{1}{k^2-1}\cdot\frac{2\cdot v_p}{T}\cdot\begin{pmatrix}\left[\cos\left(\frac{T}{2}\right)\cdot\cos\left(k\cdot\frac{T}{2}\right)+k\cdot\sin\left(\frac{T}{2}\right)\cdot\sin\left(k\cdot\frac{T}{2}\right)\right]-\\ -[\cos(i)\cdot\cos(k\cdot i)+k\cdot\sin(i)\cdot\sin(k\cdot i)]+\\ +[\cos(T)\cdot\cos(k\cdot T)+k\cdot\sin(T)\cdot\sin(k\cdot T)]-\\ -\left[\cos\left(i+\frac{T}{2}\right)\cdot\cos\left(k\cdot\left(i+\frac{T}{2}\right)\right)+k\cdot\sin\left(i+\frac{T}{2}\right)\cdot\sin\left(k\cdot\left(i+\frac{T}{2}\right)\right)\right]\end{pmatrix} \tag{7.56}$$

$$a_k = \frac{1}{k^2-1}\cdot\frac{2\cdot v_p}{T}\cdot\begin{pmatrix}[\cos(\pi)\cdot\cos(k\cdot\pi)+k\cdot\sin(\pi)\cdot\sin(k\cdot\pi)]-\\ -[\cos(i)\cdot\cos(k\cdot i)+k\cdot\sin(i)\cdot\sin(k\cdot i)]+\\ +[\cos(2\cdot\pi)\cdot\cos(2\cdot k\cdot\pi)+k\cdot\sin(2\cdot\pi)\cdot\sin(2\cdot k\cdot\pi)]-\\ -[\cos(i+\pi)\cdot\cos(k\cdot i+k\cdot\pi)+k\cdot\sin(i+\pi)\cdot\sin(k\cdot i+k\cdot\pi)]\end{pmatrix} \tag{7.57}$$

$$a_k = \frac{1}{k^2-1} \cdot \frac{2 \cdot v_p}{T} \cdot \begin{pmatrix} -1 \cdot (\mp 1) + 0 - \cos(i) \cdot \cos(k \cdot i) - k \cdot \sin(i) \cdot \sin(k \cdot i) + 1 \cdot 1 + 0 - \\ - \begin{bmatrix} -\cos(i) \cdot (\cos(k \cdot i) \cdot \cos(k \cdot \pi) - \sin(k \cdot i) \cdot \sin(k \cdot \pi)) - \\ -k \cdot \sin(i) \cdot (\sin(k \cdot i) \cdot \cos(k \cdot \pi) + \cos(k \cdot i) \cdot \sin(k \cdot \pi)) \end{bmatrix} \end{pmatrix} \tag{7.58}$$

$$a_k = \frac{1}{k^2-1} \cdot \frac{2 \cdot v_p}{T} \cdot \begin{pmatrix} \pm 1 + 1 - \cos(i) \cdot \cos(k \cdot i) - k \cdot \sin(i) \cdot \sin(k \cdot i) \mp \\ \mp \cos(i) \cdot \cos(k \cdot i) \mp k \cdot \sin(i) \cdot \sin(k \cdot i) \end{pmatrix} \tag{7.59}$$

$$a_k = \frac{1}{k^2-1} \cdot \frac{2 \cdot v_p}{T} \cdot \begin{pmatrix} \pm 1 + 1 - \cos(i) \cdot \cos(k \cdot i) \mp \cos(i) \cdot \cos(k \cdot i) - \\ - k \cdot (\sin(i) \cdot \sin(k \cdot i) \pm \sin(i) \cdot \sin(k \cdot i)) \end{pmatrix} \tag{7.60}$$

For odd k:

$$a_k = \frac{1}{k^2-1} \cdot \frac{2 \cdot v_p}{T} \cdot \begin{pmatrix} 2 - \cos(i) \cdot \cos(k \cdot i) - \cos(i) \cdot \cos(k \cdot i) - \\ -k \cdot (\sin(i) \cdot \sin(k \cdot i) + \sin(i) \cdot \sin(k \cdot i)) \end{pmatrix} \tag{7.61}$$

$$a_k = \frac{1}{k^2-1} \cdot \frac{4 \cdot v_p}{T} \cdot (1 - \cos(i) \cdot \cos(k \cdot i) - k \cdot \sin(i) \cdot \sin(k \cdot i)) \tag{7.62}$$

For even k:

$$a_k = \frac{1}{k^2-1} \cdot \frac{2 \cdot v_p}{T} \cdot \begin{pmatrix} -\cos(i) \cdot \cos(k \cdot i) + \cos(i) \cdot \cos(k \cdot i) - \\ -k \cdot (\sin(i) \cdot \sin(k \cdot i) - \sin(i) \cdot \sin(k \cdot i)) \end{pmatrix} \tag{7.63}$$

$$a_k = 0 \tag{7.64}$$

In particular, for k=0: $a_0 = 0$

Determine b_k as a function of i:

$$b_k = \frac{2}{T} \cdot \left(\int_i^{\frac{T}{2}} v_p \cdot \sin(t) \cdot \sin(k \cdot t) dt + \int_{i+\frac{T}{2}}^{T} v_p \cdot \sin(t) \cdot \sin(k \cdot t) dt \right) \tag{7.65}$$

$$b_k = \frac{2 \cdot v_p}{T} \cdot \left(\int_i^{\frac{T}{2}} \sin(t) \cdot \sin(k \cdot t) dt + \int_{i+\frac{T}{2}}^{T} \sin(t) \cdot \sin(k \cdot t) dt \right) \tag{7.66}$$

Recall [1]:

$$\int u' \cdot v = u \cdot v - \int u \cdot v' \tag{7.67}$$

$$\int \sin(t)\cdot\sin(k\cdot t)dt = -\cos(t)\cdot\sin(k\cdot t) + k\cdot\int \cos(t)\cdot\cos(k\cdot t)dt \tag{7.68}$$

$$\begin{gathered}\int \sin(t)\cdot\sin(k\cdot t)dt = -\cos(t)\cdot\sin(k\cdot t)+\\ +k\cdot\left[\sin(t)\cdot\cos(k\cdot t) + k\cdot\int \sin(t)\cdot\sin(k\cdot t)dt\right]\end{gathered} \tag{7.69}$$

$$\begin{gathered}\int \sin(t)\cdot\sin(k\cdot t)dt = -\cos(t)\cdot\sin(k\cdot t)+\\ +k\cdot\sin(t)\cdot\cos(k\cdot t) + k^2\cdot\int \sin(t)\cdot\sin(k\cdot t)dt\end{gathered} \tag{7.70}$$

$$\begin{gathered}\left(1-k^2\right)\cdot\int \sin(t)\cdot\sin(k\cdot t)dt =\\ = -\cos(t)\cdot\sin(k\cdot t) + k\cdot\sin(t)\cdot\cos(k\cdot t)\end{gathered} \tag{7.71}$$

$$\int \sin(t)\cdot\sin(k\cdot t)dt = \frac{1}{\left(1-k^2\right)}\cdot(-\cos(t)\cdot\sin(k\cdot t) + k\cdot\sin(t)\cdot\cos(k\cdot t)) \tag{7.72}$$

$$\int \sin(t)\cdot\sin(k\cdot t)dt = \frac{1}{1-k^2}\cdot(k\cdot\sin(t)\cdot\cos(k\cdot t) - \cos(t)\cdot\sin(k\cdot t)) \tag{7.73}$$

$$b_k = \frac{1}{1-k^2}\cdot\frac{2\cdot v_p}{T}\cdot\begin{pmatrix}[k\cdot\sin(t)\cdot\cos(k\cdot t) - \cos(t)\cdot\sin(k\cdot t)]_i^{\frac{T}{2}} +\\ +[k\cdot\sin(t)\cdot\cos(k\cdot t) - \cos(t)\cdot\sin(k\cdot t)]_{i+\frac{T}{2}}^{T}\end{pmatrix} \tag{7.74}$$

$$b_k = \frac{1}{1-k^2}\cdot\frac{2\cdot v_p}{T}\cdot\begin{pmatrix}\left[k\cdot\sin\left(\frac{T}{2}\right)\cdot\cos\left(k\cdot\frac{T}{2}\right) - \cos\left(\frac{T}{2}\right)\cdot\sin\left(k\cdot\frac{T}{2}\right)\right]-\\ -[k\cdot\sin(i)\cdot\cos(k\cdot i) - \cos(i)\cdot\sin(k\cdot i)]+\\ +[k\cdot\sin(T)\cdot\cos(k\cdot T) - \cos(T)\cdot\sin(k\cdot T)]-\\ -\left[k\cdot\sin\left(i+\frac{T}{2}\right)\cdot\cos\left(k\cdot\left(i+\frac{T}{2}\right)\right) - \cos\left(i+\frac{T}{2}\right)\cdot\sin\left(k\cdot\left(i+\frac{T}{2}\right)\right)\right]\end{pmatrix} \tag{7.75}$$

$$b_k = \frac{1}{1-k^2}\cdot\frac{2\cdot v_p}{T}\cdot\begin{pmatrix}0 - k\cdot\sin(i)\cdot\cos(k\cdot i) + \cos(i)\cdot\sin(k\cdot i) + 0+\\ +k\cdot\sin(i)\cdot\cos(k\cdot i + k\cdot\pi) - \cos(i)\cdot\sin(k\cdot i + k\cdot\pi)\end{pmatrix} \tag{7.76}$$

$$b_k = \frac{1}{1-k^2}\cdot\frac{2\cdot v_p}{T}\cdot\begin{pmatrix}-k\cdot\sin(i)\cdot\cos(k\cdot i) + \cos(i)\cdot\sin(k\cdot i)+\\ +k\cdot\sin(i)\cdot(\cos(k\cdot i)\cdot\cos(k\cdot\pi) - \sin(k\cdot i)\cdot\sin(k\cdot\pi))-\\ -\cos(i)\cdot(\sin(k\cdot i)\cdot\cos(k\cdot\pi) + \cos(k\cdot i)\cdot\sin(k\cdot\pi))\end{pmatrix} \tag{7.77}$$

$$b_k = \frac{1}{1-k^2} \cdot \frac{2 \cdot v_p}{T} \cdot (\pm \cos(i) \cdot \sin(k \cdot i) + \cos(i) \cdot \sin(k \cdot i) - k \cdot \sin(i) \cdot \cos(k \cdot i) \mp k \cdot \sin(i) \cdot \cos(k \cdot i)) \tag{7.78}$$

$$b_k = \frac{1}{k^2-1} \cdot \frac{2 \cdot v_p}{T} \cdot (\mp \cos(i) \cdot \sin(k \cdot i) - \cos(i) \cdot \sin(k \cdot i) + k \cdot \sin(i) \cdot \cos(k \cdot i) \pm k \cdot \sin(i) \cdot \cos(k \cdot i)) \tag{7.79}$$

For odd k:

$$b_k = \frac{1}{k^2-1} \cdot \frac{4 \cdot v_p}{T} \cdot (-\cos(i) \cdot \sin(k \cdot i) + k \cdot \sin(i) \cdot \cos(k \cdot i)) \tag{7.80}$$

For even k:

$$b_k = 0 \tag{7.81}$$

The function for the phase-controlled voltage v as a function of time t is (k odd)

$$v(t) = \sum_{k=1}^{\infty} (a_k \cdot \cos(k \cdot t) + b_k \cdot \sin(k \cdot t)) \tag{7.82}$$

$$v(t) = \sum_{k=1}^{\infty} \left(\left(\frac{1}{k^2-1} \cdot \frac{4 \cdot v_p}{T} \cdot (1 - \cos(i) \cdot \cos(k \cdot i) - k \cdot \sin(i) \cdot \sin(k \cdot i)) \right) \cdot \cos(k \cdot t) - \left(\frac{1}{k^2-1} \cdot \frac{4 \cdot v_p}{T} \cdot (\cos(i) \cdot \sin(k \cdot i) - k \cdot \sin(i) \cdot \cos(k \cdot i)) \right) \cdot \sin(k \cdot t) \right) \tag{7.83}$$

$$v(t) = \frac{4 \cdot v_p}{T} \cdot \sum_{k=1}^{\infty} \frac{(1 - \cos(i) \cdot \cos(k \cdot i) - k \cdot \sin(i) \cdot \sin(k \cdot i)) \cdot \cos(k \cdot t) - (\cos(i) \cdot \sin(k \cdot i) - k \cdot \sin(i) \cdot \cos(k \cdot i)) \cdot \sin(k \cdot t)}{k^2-1} \tag{7.84}$$

$$v(t) = \frac{4 \cdot v_p}{T} \cdot \sum_{k=1}^{\infty} \frac{1 \cdot \cos(k \cdot t) - \cos(i) \cdot \cos(k \cdot i) \cdot \cos(k \cdot t) - k \cdot \sin(i) \cdot \sin(k \cdot i) \cdot \cos(k \cdot t) - \cos(i) \cdot \sin(k \cdot i) \cdot \sin(k \cdot t) + k \cdot \sin(i) \cdot \cos(k \cdot i) \cdot \sin(k \cdot t)}{k^2-1} \tag{7.85}$$

$$v(t) = \frac{4 \cdot v_p}{T} \cdot \sum_{k=1}^{\infty} \frac{\cos(k \cdot t) - \cos(i) \cdot (\cos(k \cdot i) \cdot \cos(k \cdot t) + \sin(k \cdot i) \cdot \sin(k \cdot t)) - k \cdot \sin(i) \cdot (\sin(k \cdot i) \cdot \cos(k \cdot t) - \cos(k \cdot i) \cdot \sin(k \cdot t))}{k^2-1} \tag{7.86}$$

$$v(t) = \frac{4 \cdot v_p}{T} \cdot \sum_{k=1}^{\infty} \frac{\cos(k \cdot t) - \cos(i) \cdot \cos(k \cdot (i-t)) - k \cdot \sin(i) \cdot \sin(k \cdot (i-t))}{k^2-1} \tag{7.87}$$

For k = 1, a problem may arise. Therefore, the limit for k → 1 must be considered:

$$\lim_{k\to 1}\frac{\cos(k\cdot t)-\cos(i)\cdot\cos(k\cdot(i-t))-k\cdot\sin(i)\cdot\sin(k\cdot(i-t))}{k^2-1}= \quad (7.88a)$$

This term is of the type 0/0, so l'Hospital's rule can be applied:

$$\lim_{k\to 1}\frac{-t\cdot\sin(k\cdot t)+(i-t)\cdot\cos(i)\cdot\sin(k\cdot(i-t))-\sin(i)\cdot\sin(k\cdot(i-t))-k\cdot(i-t)\cdot\sin(i)\cdot\cos(k\cdot(i-t)}{2\cdot k} \quad (7.88b)$$

$$=\frac{-t\cdot\sin(t)+(i-t)\cdot\cos(i)\cdot\sin(i-t)-\sin(i)\cdot\sin(i-t)-(i-t)\cdot\sin(i)\cdot\cos(i-t)}{2}= \quad (7.88c)$$

$$=\frac{-t\cdot\sin(t)-(i-t)\cdot\sin(t)-\sin(i)\cdot\sin(i-t)}{2}= \quad (7.88d)$$

$$=\frac{-i\cdot\sin(t)-\sin(i)\cdot\sin(i-t)}{2} \quad (7.88e)$$

$$v(t)=\frac{4\cdot v_p}{T}\cdot\left(\begin{array}{c}\frac{-i\cdot\sin(t)-\sin(i)\cdot\sin(i-t)}{2}+\\ +\sum_{k=2}^{\infty}\frac{\cos(k\cdot t)-\cos(i)\cdot\cos(k\cdot(i-t))-k\cdot\sin(i)\cdot\sin(k\cdot(i-t))}{k^2-1}\end{array}\right) \quad (7.89)$$

with i as the parameter for the switching time of the thyristors. Since only odd values of k are relevant, k is substituted as follows:

$$k\to 2\cdot k+1 \quad (7.90)$$

$$v(t)=\frac{4\cdot v_p}{T}\cdot\left(\begin{array}{c}\frac{-i\cdot\sin(t)-\sin(i)\cdot\sin(i-t)}{2}+\\ \cos((2\cdot k+1)\cdot t)-\cos(i)\cdot\cos((2\cdot k+1)\cdot(i-t))-\\ +\sum_{k=1}^{\infty}\frac{-(2\cdot k+1)\cdot\sin(i)\cdot\sin((2\cdot k+1)\cdot(i-t))}{(2\cdot k+1)^2-1}\end{array}\right) \quad (7.91)$$

If this Fourier series is considered for several values of i, then for a set amplitude of $v_p = 5$ and i = 0.2, 0.5, and 0.8 in Excel (with the first 20 harmonics displayed), the results are shown in Figs. 7.6, 7.7, and 7.8:

For i in the range of 0 (thyristor fully turned off) or 1 (thyristor fully turned on), even more interesting phenomena arise (Figs. 7.9, 7.10, 7.11, and 7.12).

In all cases, attention should also be paid to the amplitude.

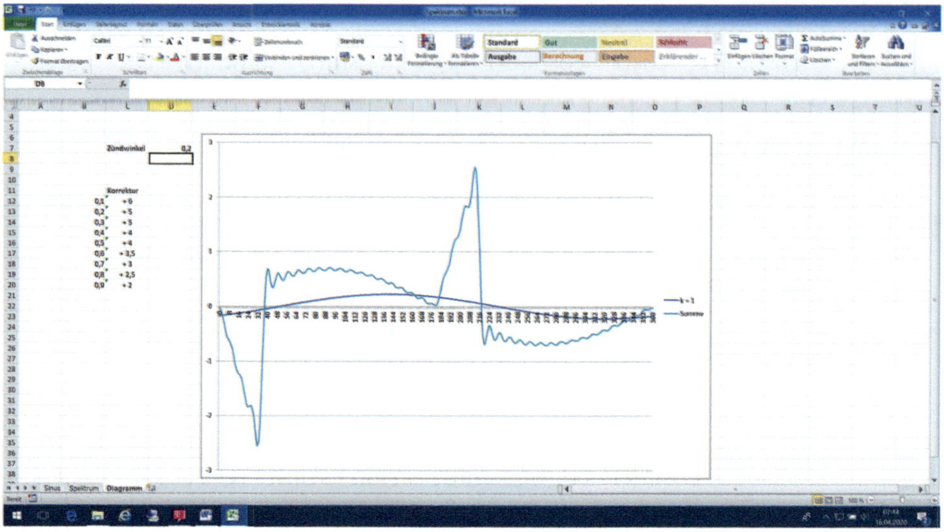

Fig. 7.6 Graph of the sum of the Fourier series up to the 20th harmonic of a phase-controlled sine wave with a control factor i = 0.2

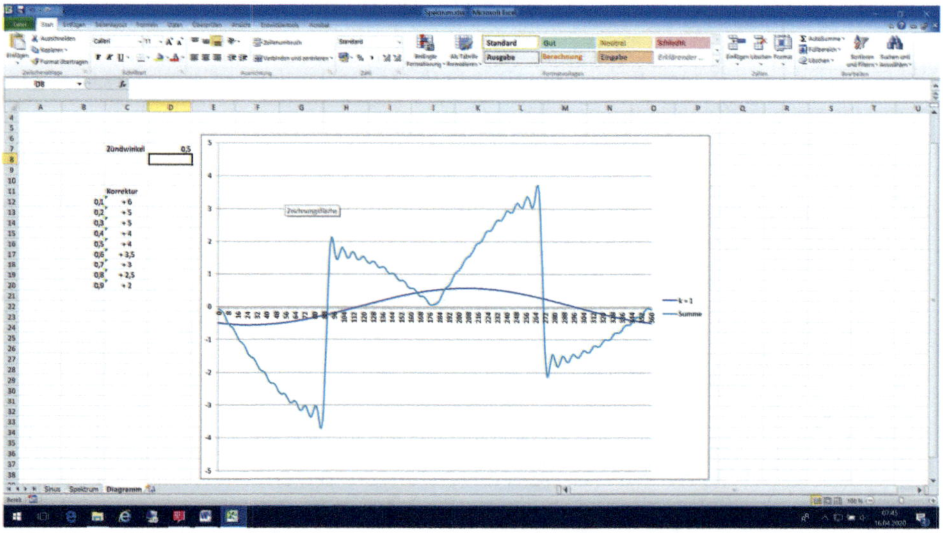

Fig. 7.7 Graph of the sum of the Fourier series up to the 20th harmonic of a phase-controlled sine wave with a control factor i = 0.5

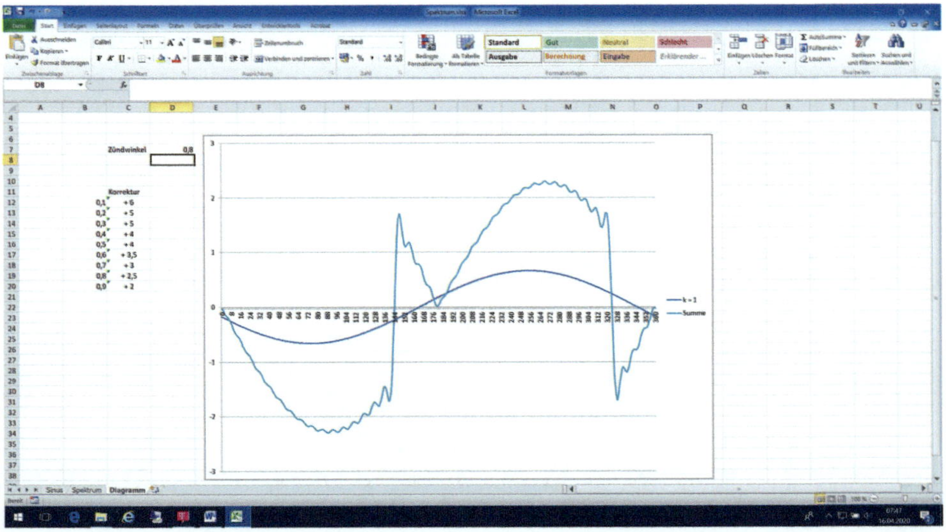

Fig. 7.8 Graph of the sum of the Fourier series up to the 20th harmonic of a phase-controlled sine wave with a control factor i = 0.8

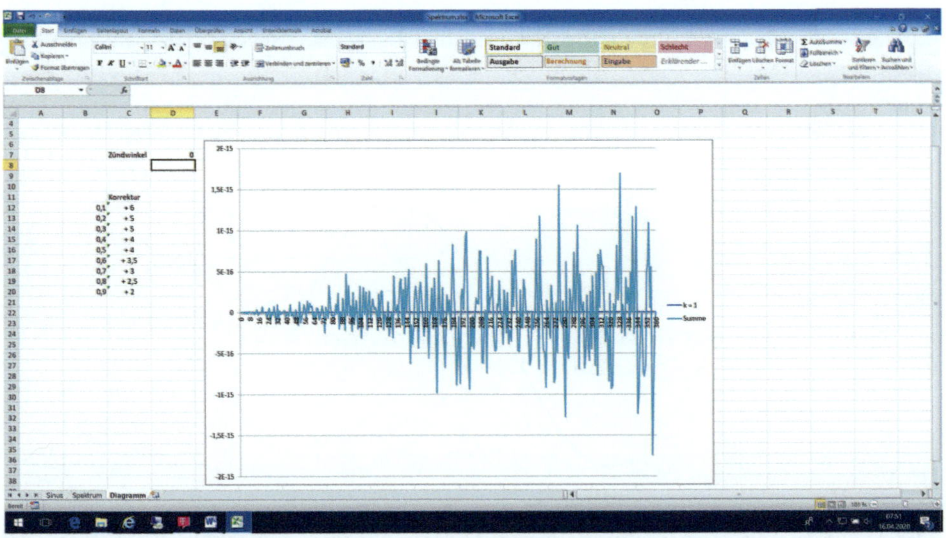

Fig. 7.9 Graph of the sum of the Fourier series up to the 20th harmonic of a phase-controlled sine signal with a control factor i = 0

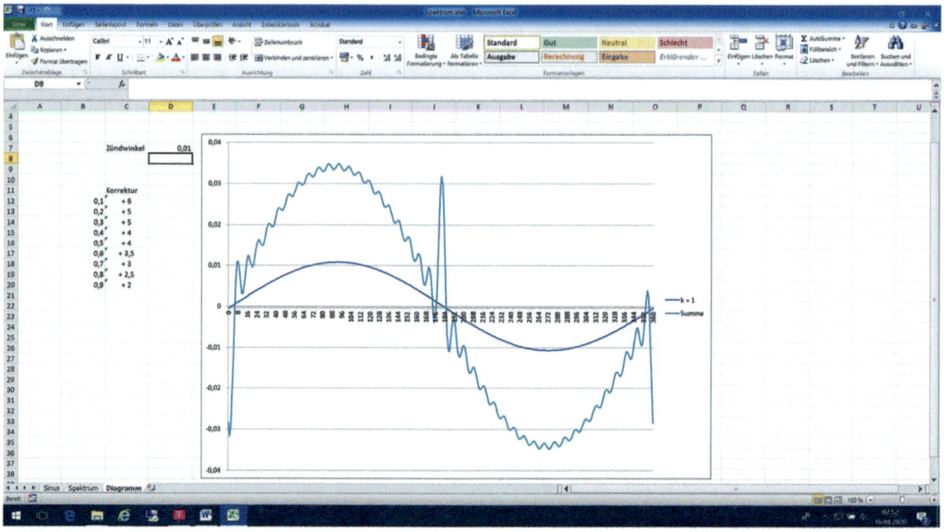

Fig. 7.10 Graph of the sum of the Fourier series up to the 20th harmonic of a phase-controlled sine signal with a control factor i = 0.01

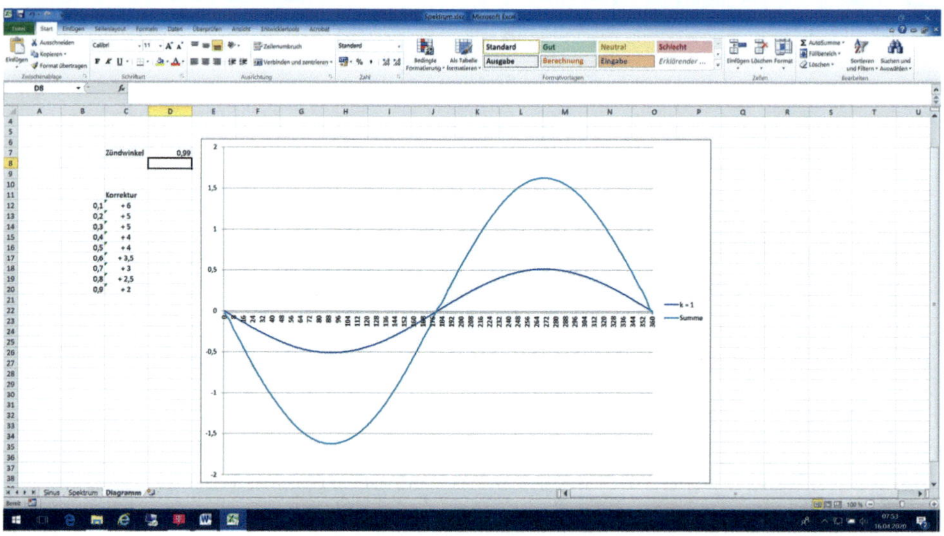

Fig. 7.11 Graph of the sum of the Fourier series up to the 20th harmonic of a phase-controlled sine signal with a control factor i = 0.99

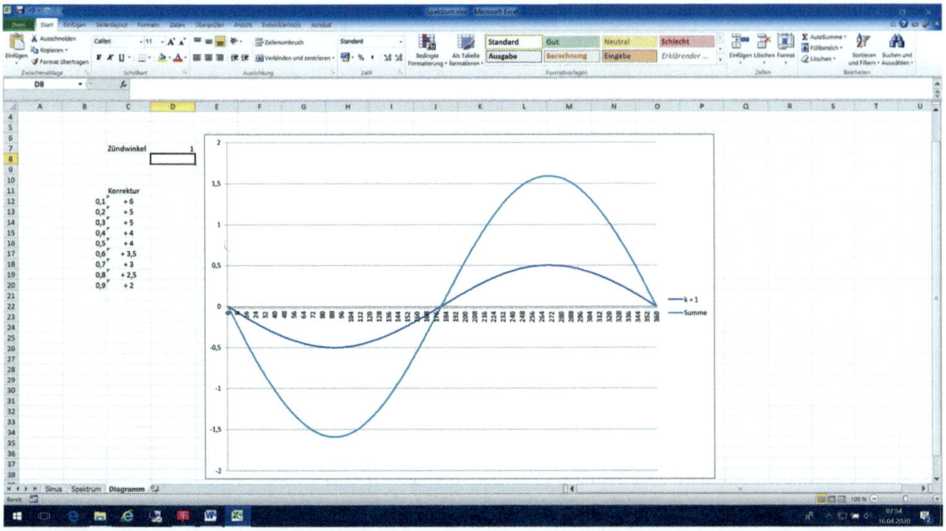

Fig. 7.12 Graph of the sum of the Fourier series up to the 20th harmonic of a phase-controlled sine signal with a control factor i = 1

Additionally, when using Excel for visualization, it is important to ensure that the arguments in the trigonometric functions are normalized to $2 \cdot \pi$.

I have checked the calculation of the Fourier series many times and can no longer find any deficiencies. However, in [2] I found some statements regarding the convergence of the Fourier series, including the following:“ Although one can readily construct a Fourier series for a periodic function, the series does not necessarily converge …”.

References

1. Bronstein, Semendjajew; Taschenbuch der Mathematik; Verlag Nauka Moskau, 25. Auflage 1991
2. Wikipedia, Fourierreihe, 16. 4. 2020

The Diode Equation

8

I engaged with the diode equation because I needed to determine the losses of a power rectifier as a function of temperature within the power rectifier. The approach seemed straightforward: I would use the Shockley equation, determine the two factors for the diode based on the characteristic curve at 20 °C, and then use temperature as a parameter on which the power loss depends. In practice, this did not work, because a diode is not as simple as the Shockley equation suggests, and because a diode exhibits a temperature behavior that is contrary to what is described by the Shockley equation.

The operating principle of a diode describes an electrical component that is essentially low-resistance in one voltage direction, allowing a significant electric current to flow, and in the opposite direction blocks the electric current due to its high-resistance property. This is why diodes were often referred to as "valves" in historical contexts. Furthermore, diodes are characterized by highly nonlinear behavior, meaning that the ratio of an infinitesimal voltage change to an infinitesimal current change is by no means constant:

$$\frac{dU_{Diode}}{dI_{Diode}} \neq \text{const} \tag{8.1}$$

The valve function of a diode is technologically realized by two principles:

- By the emission of charges from a conductor into a gas or into a vacuum, and
- by the passage of charge carriers through a transition layer in a semiconductor crystal,

Strictly speaking, "the semiconductor crystal" is a misnomer, because the transition layer is actually located between differently structured (semiconductor) crystals, and thus, as charge carriers move from one structure to another, they encounter new energetic conditions—similar to the situation when a charge carrier leaves a conductor and enters a gas or vacuum [1; 2].

J. von Stackelberg, *The Mass of a Photon*, https://doi.org/10.1007/978-3-658-49408-7_8

8.1 The Vacuum Diode

In the vacuum diode, the rectifying effect is achieved by heating one electrode, namely the cathode, to facilitate the emission of electrons or to enable the emission of a significant number of electrons. When a voltage is applied between the cathode and the anode, the emitted electrons move as a charge carrier current toward the anode. The emission behavior of the vacuum diode is described by the Richardson equation [3]:

$$J = A \cdot T^2 \cdot e^{-\frac{W_e}{k_B \cdot T}} \tag{8.2}$$

with

$J=$	Current density of the emitted electrons in A/m^2
$A=$	Richardson constant $= 1.20173 \cdot 10^6$ $A/(m^2 \cdot K^2)$, but it also appears to be material-dependent
$T=$	absolute temperature in K
$W_e=$	Work function for electrons in J (approx. 1 … 6 eV, where 1 eV $= 1.602 \cdot 10^{-19}$ J) [2]
$k_B=$	Boltzmann constant $= 1.38064853 \cdot 10^{-23}$ J/K

Applying an electric voltage between the cathode and anode causes current to flow. The rectifying effect arises primarily because almost exclusively electrons can be emitted as charge carriers from the cathode.

A similar effect can also be observed in strong electric field configurations without thermionic emission, as found, for example, in electrostatic precipitators. This effect is described by the so-called Fowler–Nordheim equation [4; 5].

8.2 The Semiconductor Diode

Whereas in the vacuum diode the rectifying interface is the physical boundary between a material and empty space, in a semiconductor diode this interface is located at the boundary layer between two different material crystal structures.

Semiconductors as materials have electrical conductivity between that of conductors and insulators, although the boundaries are somewhat arbitrary in terms of values [6]. The electrical conductivity of a semiconductor crystal is relatively low and highly temperature-dependent (Fig. 8.1).

The energy required to free an electron from the atomic bond and make it a conduction electron (transition from the valence band to the conduction band) therefore depends on temperature (Fig. 8.2).

To increase the conductivity of a semiconductor crystal, it is "doped." Technically, this involves introducing foreign atoms into the crystal, which create either an excess of electrons (n-type doping) or a deficiency of electrons (p-type doping) in the lattice, thereby providing mobile charge carriers.

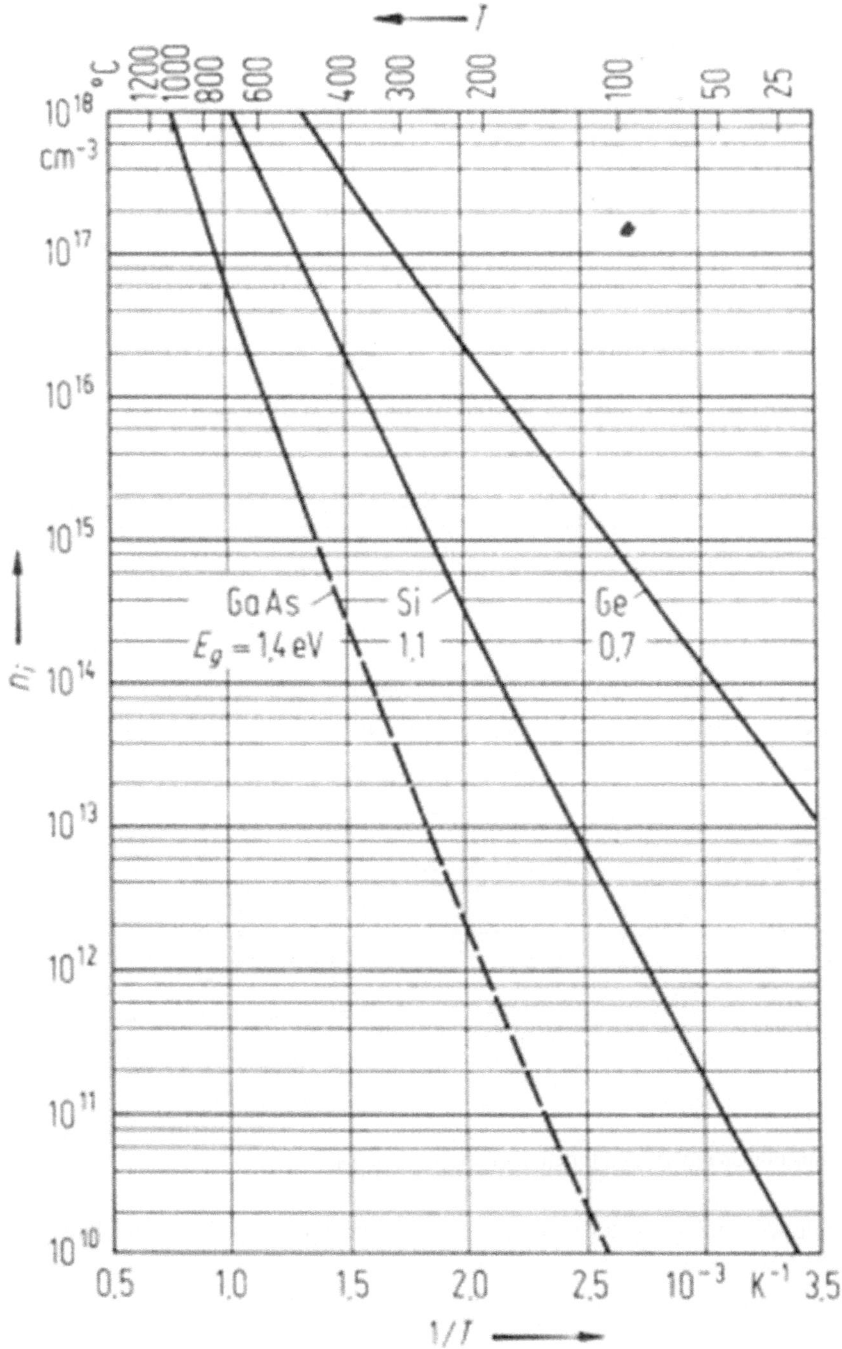

Fig. 8.1 Representation of the number of free electrons per cm^3 (n_i on the left ordinate) as a function of temperature T (upper abscissa) for undoped GaAs, Si, and Ge [7]

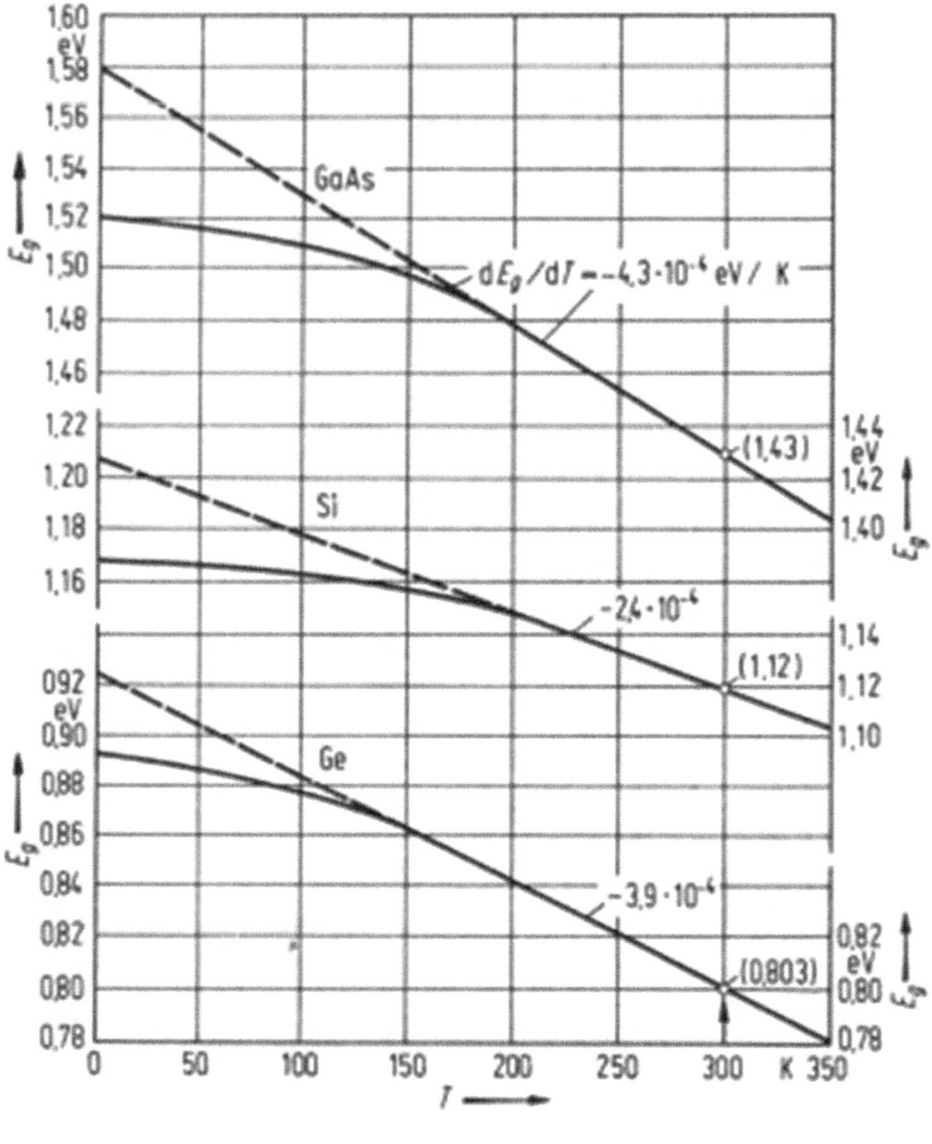

Fig. 8.2 Transition energy E_g as a function of temperature T [8]

Even in doped semiconductor material, a thermal dependence of the free charge carriers can still be observed (Figs. 8.3 and 8.4).

In doped material, a distinction is made between so-called majority carriers, which are the free charge carriers generated by doping, and minority carriers, which are the charge carriers of the opposite polarity. For example, in a silicon crystal, which essentially consists of tetravalent silicon and is doped with pentavalent phosphorus, the free

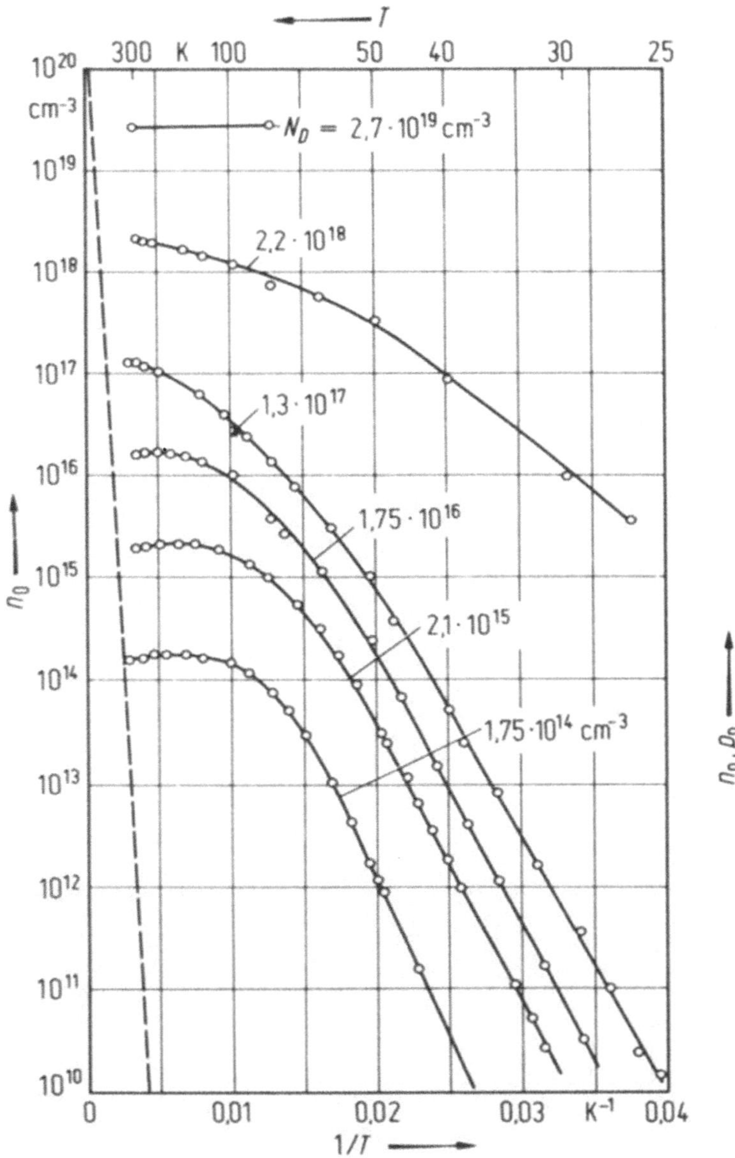

Fig. 8.3 Number of free charge carriers as a function of temperature [9]

electrons are the majority carriers and the holes are the minority carriers, since each phosphorus dopant atom in the crystal lattice provides a fifth electron.

For majority carriers, a saturation of the free charge carrier concentration can be observed at a certain temperature with respect to their thermal dependence, whereas minority carriers are increasingly released at higher temperatures (Fig. 8.5).

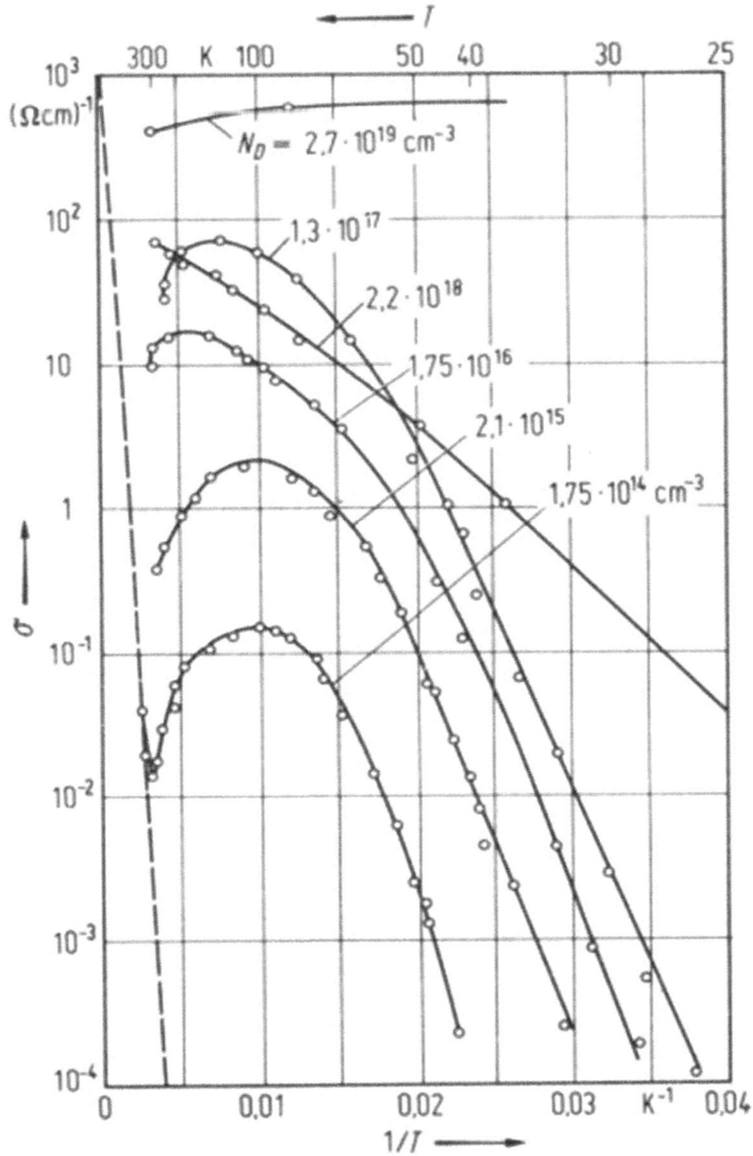

Fig. 8.4 Number of free charge carriers as a function of temperature [10]

When an electric field is applied to the semiconductor, the charge carriers are set in motion. The velocity of this movement—the drift velocity—depends on the field strength up to a certain point and then reaches saturation (Fig. 8.6).

As doping increases, the mobility of charge carriers in the crystal lattice decreases due to the increase in lattice disturbances (Fig. 8.7).

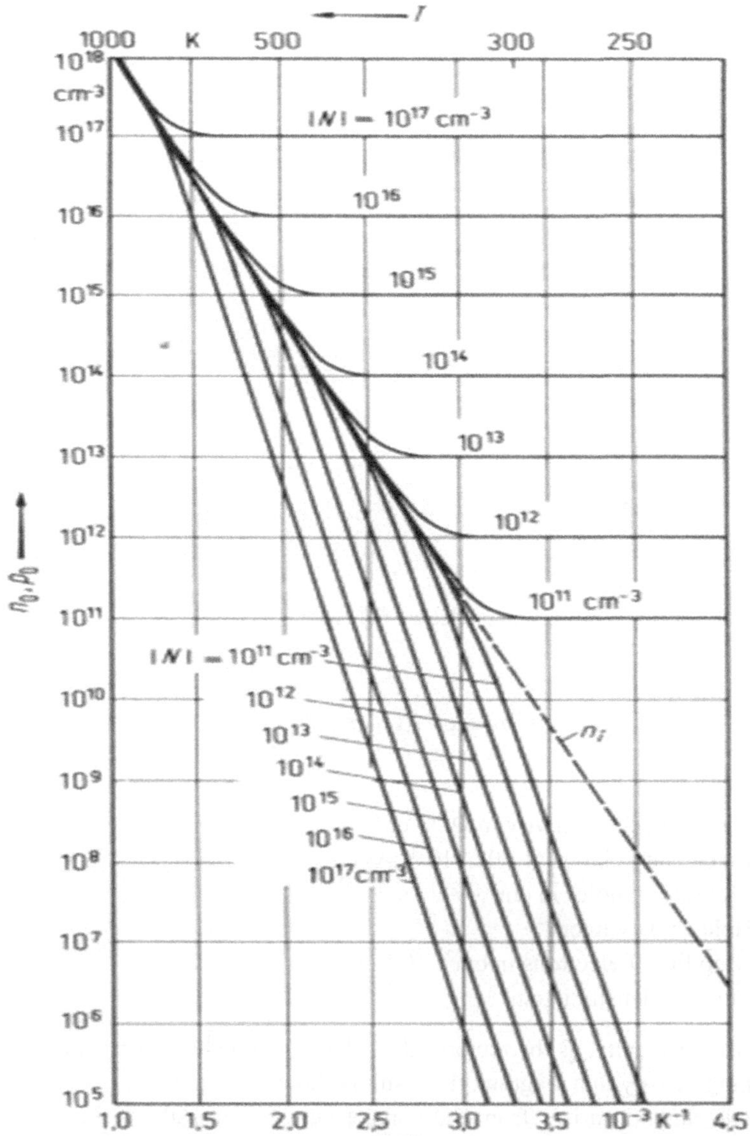

Fig. 8.5 Free charge carriers as a function of temperature, showing saturation for majority carriers [11]

In addition, the increasing thermal motion of atoms within the crystal lattice with rising temperature leads to a reduction in the mobility of charge carriers (Figs. 8.8 and 8.9).

The electrical conductivity of a semiconductor material is calculated as the product of charge carrier density and charge carrier mobility:

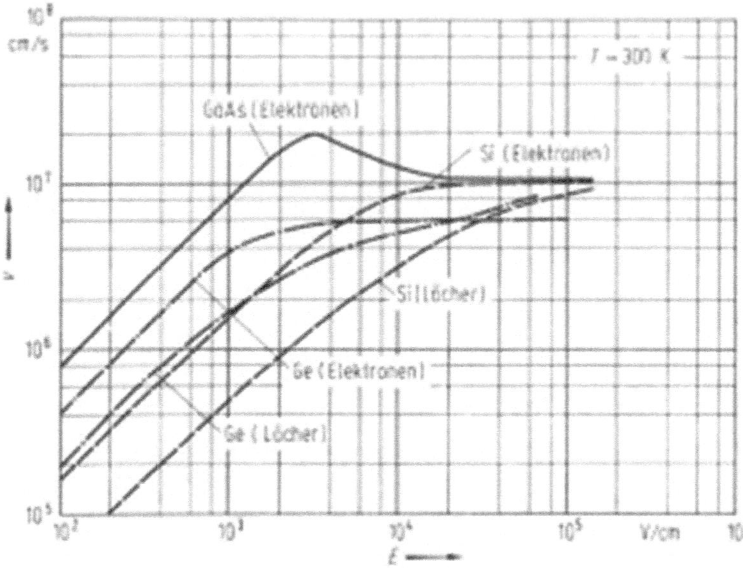

Fig. 8.6 Drift velocity as a function of electric field strength [12]

$$\sigma = e \cdot (\mu_p \cdot p + \mu_n \cdot n) \tag{8.3}$$

with

$\sigma =$	Electrical conductivity in $\Omega^{-1} \cdot cm^{-1}$
$e =$	Elementary charge ($1.602 \cdot 10^{-19}$ As)
$\mu_p =$	Mobility of holes in $cm^2 \cdot V^{-1} \cdot s^{-1}$
$p =$	Hole density in cm^{-3}
$\mu_n =$	Mobility of electrons in $cm^2 \cdot V^{-1} \cdot s^{-1}$
$n =$	Electron density in cm^{-3}

If the specific resistivity, as the reciprocal of electrical conductivity, is plotted as a function of doping density in a diagram, the result is shown in Fig. 8.10.

The diode function arises from the contact between crystals of different structures. These may be semiconductor crystals with different doping (n-doped and p-doped) or semiconductors and metals, as in Schottky diodes.

Specifically, the contact between different crystal structures causes charge carriers to migrate from one crystal to the other at the junction, in the absence of an external electric field, i.e., at equilibrium. For example, electrons migrate from n-doped silicon into p-doped silicon to fill the positions of holes. In this way, a narrow boundary layer forms in the transition region, in which the outermost electron shells correspond to those of the pure semiconductor; this transition region, where the charge carriers have balanced out,

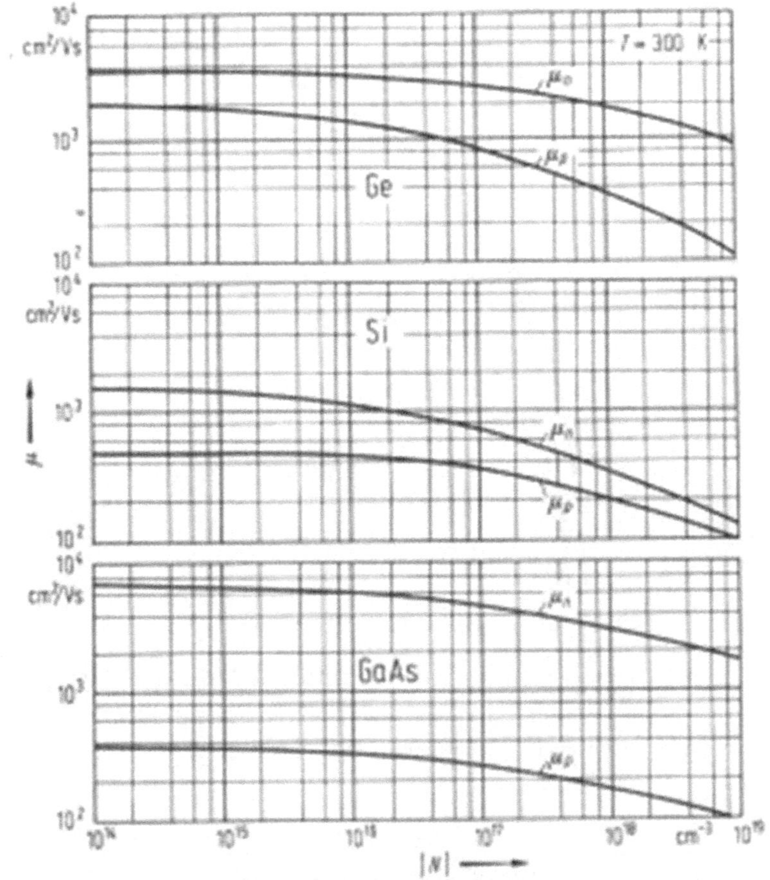

Fig. 8.7 Mobility as a function of doping concentration [13]

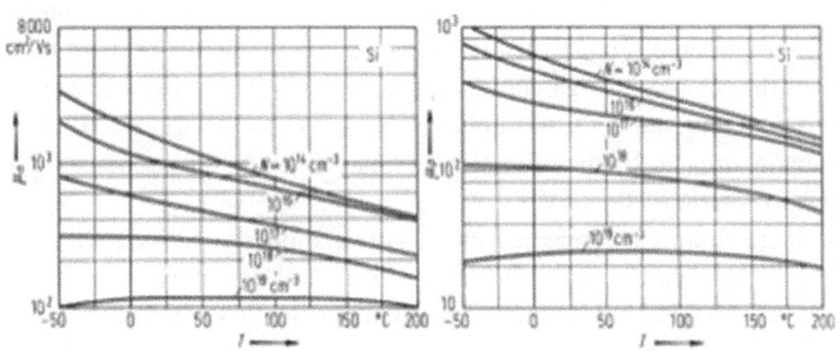

Fig. 8.8 Temperature dependence of drift mobility [14]

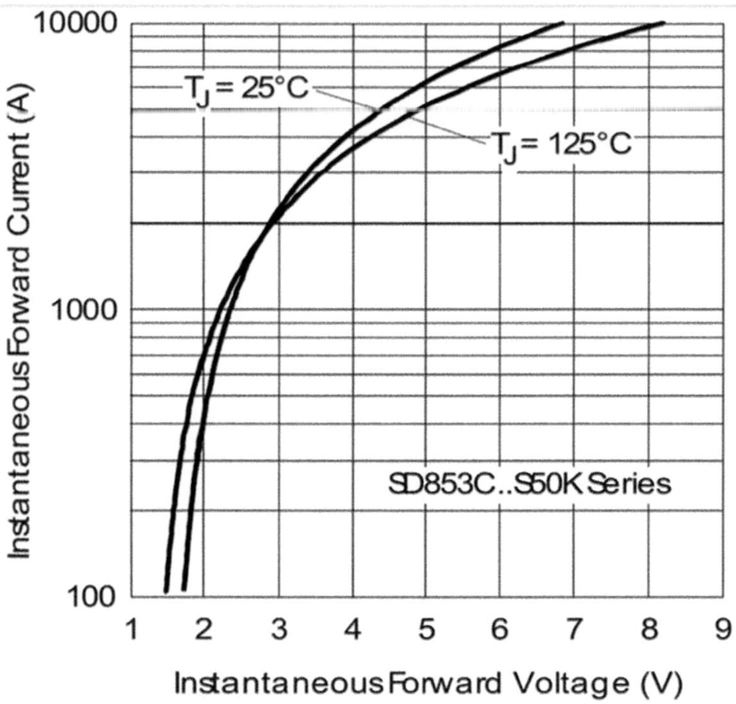

Fig. 8.9 Temperature dependence of drift mobility [15]

is called the space charge region. The electrical resistance of this narrow boundary layer is equivalent to that of an intrinsic semiconductor. The migration of charge carriers in the transition region creates an electric potential, because although the atoms have adopted a semiconductor configuration in their outer shells, the number of electrons and protons is no longer balanced (Fig. 8.11).

When an electric voltage is applied in the "forward direction," i.e., the negative terminal of the voltage source is connected to the n-doped side and the positive terminal to the p-doped side, the missing charge carriers (electrons or holes) are gradually compensated depending on the magnitude of the applied voltage. When the applied voltage equals the electric potential created by the migration of charge carriers in the transition region, the diode begins to conduct.

If the voltage source is connected in the "reverse direction," i.e., with the polarities reversed, the space charge region gradually widens until the entire semiconductor crystal of the diode is in the state of an intrinsic semiconductor. This means that as the reverse voltage increases, the resistance of the diode becomes ever higher, and the applied voltage increases, resulting in the reverse current through the diode remaining nearly constant in certain regions.

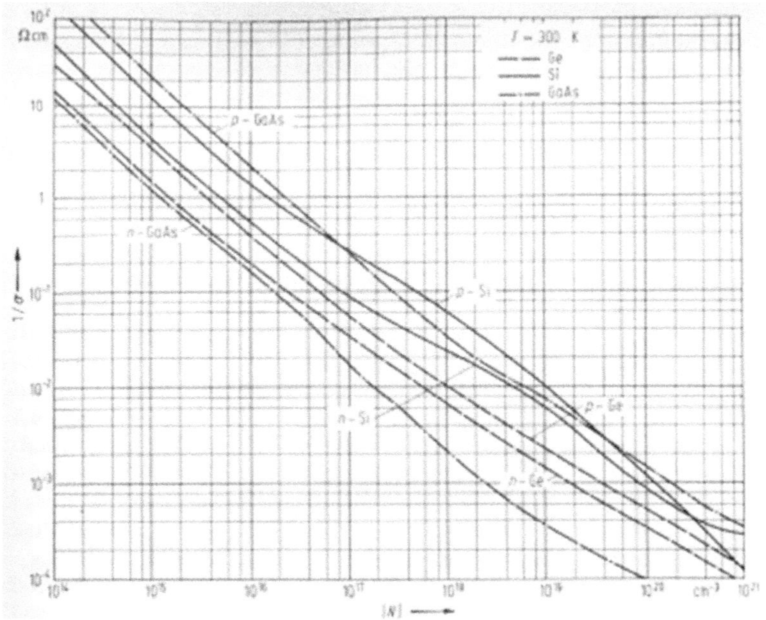

Fig. 8.10 Specific resistivity as a function of doping density [16]

If the reverse voltage is further increased, an avalanche breakdown eventually occurs in the semiconductor crystal, causing the diode to become low-resistance; the voltage across the semiconductor crystal remains nearly constant, regardless of the current magnitude. This effect is utilized, for example, in so-called Zener diodes. For the sake of completeness, it should be mentioned that the operation of Zener diodes is based on two effects: the Zener effect and the avalanche effect.

To mathematically model the electrical behavior of semiconductor diodes, William B. Shockley introduced an equation that describes the forward current of a diode as a function of the reverse saturation current, the applied voltage, and the temperature [18]:

$$I_D = I_S(T) \cdot \left(e^{\frac{e \cdot U}{n \cdot k \cdot T}} - 1\right) \tag{8.4}$$

with

$I_D =$	Diode forward current in A
$I_S =$	Diode reverse saturation current (10^{-12} … 10^{-6} A)
$e =$	Elementary charge (1.602×10^{-19} C)
$U =$	Diode voltage in V
$n =$	Diode-specific constant (1 … 2)
$k =$	Boltzmann constant ($1.38064853 \times 10^{-23}$ J/K)
$T =$	Absolute temperature in K

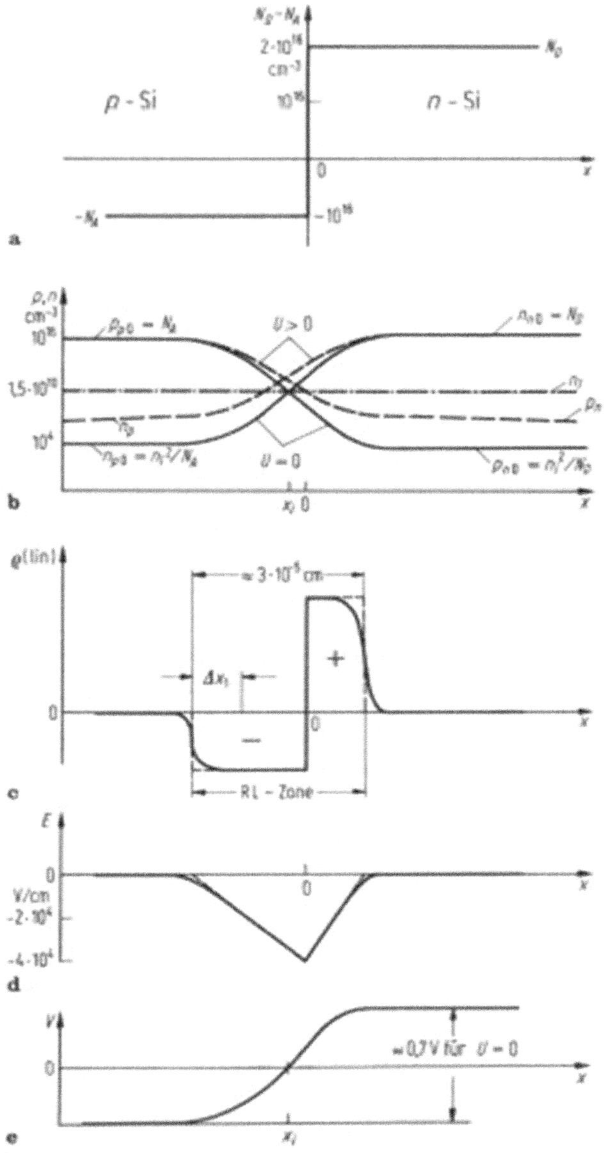

Fig. 8.11 Migration effect at the junction of different semiconductor layers [17]

This diode equation, which is well established in the literature, has a few shortcomings:

- It does not account for the negative signs of the reverse saturation current and the elementary charge of the electron

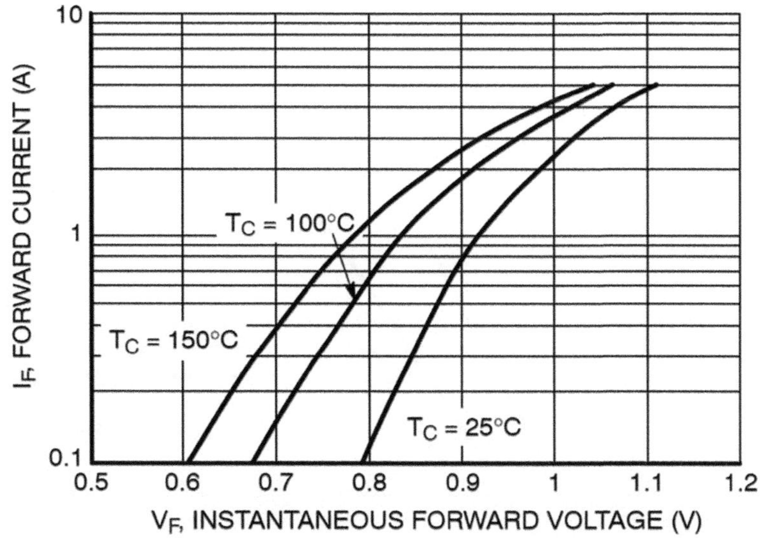

Fig. 8.12 Example of a family of characteristic curves for a semiconductor diode with temperature as a parameter [19]

- The reverse currents specified in diode datasheets are usually several orders of magnitude higher than the stated range of $10^{-12} \ldots 10^{-6}$ A
- It only describes the actual behavior of the diode over very small voltage ranges [18]
- If, for a fixed value of n, different temperatures are used as parameters for a family of characteristic curves, the resulting behavior is opposite to that of a real diode; that is, in a real diode, the voltage required for a given current decreases as temperature increases, whereas in the Shockley equation, the voltage required for a given current increases with temperature (see Figs. 8.12 and 8.13).

My concept for a mathematical model of a diode equation is to adjust certain parameters of the diode equation based on the characteristic curves of a real diode, which can be found, for example, in datasheets. This adjusted diode equation should then be sufficiently accurate over a wider operating range, such as the active forward region, for the three variables voltage, current, and temperature, to enable calculations or simulations. Furthermore, values such as the reverse current should also be obtainable from the datasheet.

Essentially, the forward behavior of a semiconductor diode can be described using an exponential function. However, to adequately describe the temperature dependence, the temperature must be included in the numerator of the exponent in the exponential function.

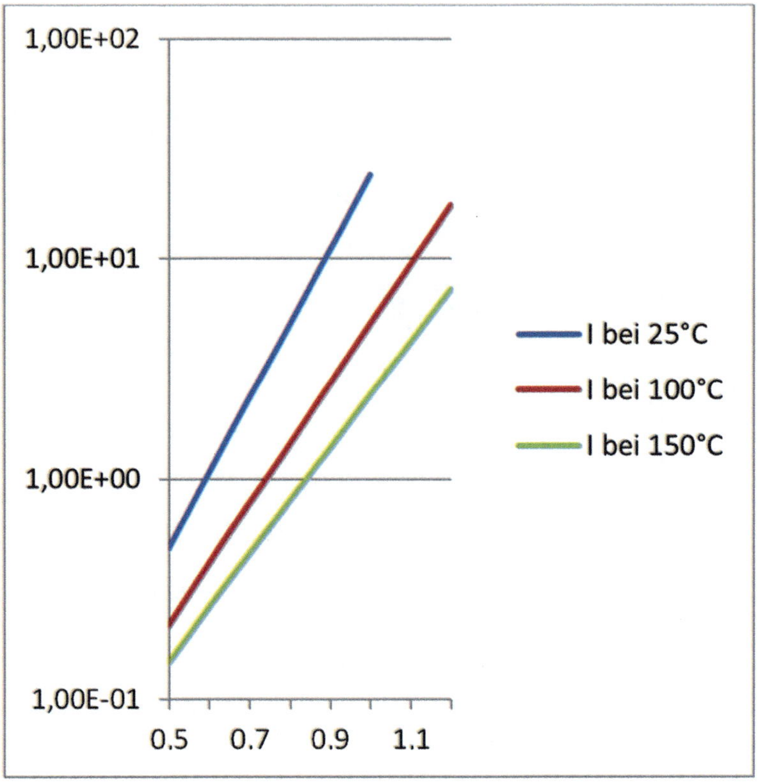

Fig. 8.13 Example of a diode characteristic field based on the Shockley equation

Similar to the Richardson equation, which, among other things, incorporates the work function of electrons leaving the cathode, the ionization energy could be included in the diode equation. This yields a diode equation of the form:

$$I_D = |I_S| \cdot \left(e^{\frac{q \cdot f(U)}{W_i}} - 1\right) \cdot \left(e^{\frac{k \cdot f(T)}{W_i}} - 1\right) \quad (8.5)$$

with

I_D = Diode forward current in A
I_S = Diode reverse saturation current in A
q = Elementary charge (1.602×10^{-19} C)
U = Diode voltage in V
W_i = Ionization energy in eV (1 eV = 1.602×10^{-19} J)
k = Boltzmann constant ($1.38064853 \times 10^{-23}$ J/K)
T = Absolute temperature in K

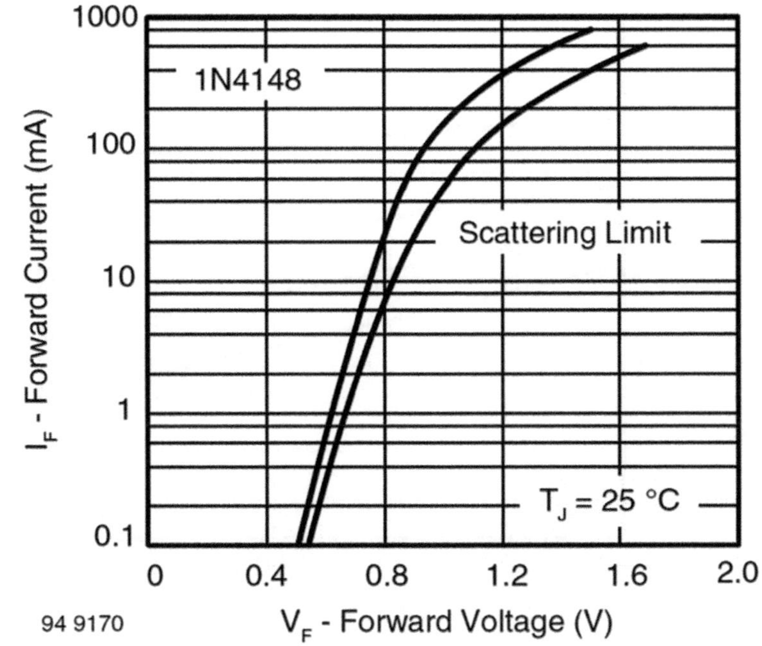

Fig. 8.14 Characteristic field of a 1N4148 [20]

For f(U) and f(T), functions with suitable parameters must be found that allow the general proposed diode equation to be adapted to the properties of specific real diodes.

As previously described, the forward current of a diode—apart from the Hall and photoelectric effects—depends on both temperature and the applied voltage. This dependence is nonlinear, even with respect to exponential relationships. To find an approximation function with sufficient accuracy, a certain number of supporting points are taken from the actual characteristic curve, these are processed in a suitable system of equations, and the resulting solution function is given as follows:

$$f(U) = a \cdot U^5 + b \cdot U^4 + c \cdot U^3 + d \cdot U^2 + e \cdot U + f \tag{8.6}$$

The initial approach is to use a 5th-order polynomial, since a 3rd-order polynomial allows for only four supporting points, which is certainly insufficient for a real diode characteristic (Fig. 8.14).

Furthermore, for a 4th-order or 6th-order polynomial, if the coefficient of the highest power of the argument is positive, the polynomial tends toward $+\infty$ as the argument approaches $-\infty$. For the exponential function, this means that the function value also tends toward $+\infty$, which does not correspond to the behavior of a diode.

Thus, six coefficients a … f must be determined, which corresponds to six supporting points for the polynomial (see Chap. 9).

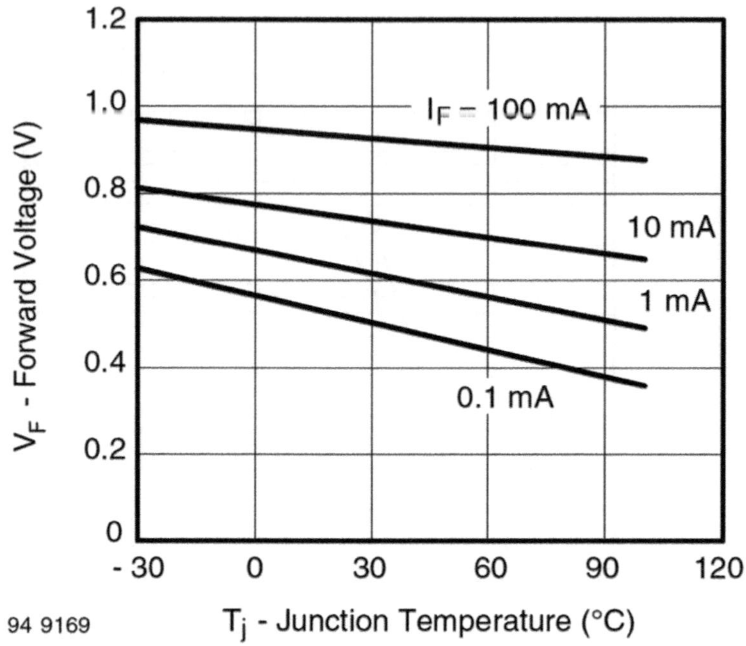

Fig. 8.15 Forward characteristics of a semiconductor diode [21]

For f(T), it must be considered that as temperature increases, there is a certain temperature at which the conductivity of the semiconductor reaches a maximum. This leads to the consideration of introducing a polynomial of even order. When examining the dependence of a diode's conductivity on temperature using several example datasheets (Figs. 8.15 and 8.16).

A 2nd-order polynomial appears to provide sufficient accuracy. Since, in terms of thermal behavior, it cannot be assumed for doped semiconductors that the conductivity is zero at a temperature of T = 0 K, all coefficients are determined, which leads to three supporting points (for determination of the coefficients, see Chap. 9).

$$\mathrm{f(T)} = \mathrm{u} \cdot \mathrm{T}^2 + \mathrm{v} \cdot \mathrm{T} + \mathrm{w} \tag{8.7}$$

This results in the diode equation:

$$I_D = |I_S| \cdot \left(e^{\frac{q \cdot \left(a \cdot U^5 + b \cdot U^4 + c \cdot U^3 + d \cdot U^2 + e \cdot U + f \right)}{W_i}} - 1 \right) \cdot \left(e^{\frac{k \cdot \left(u \cdot T^2 + v \cdot T + w \right)}{W_i}} - 1 \right) \tag{8.8}$$

With regard to the units for the coefficients a … f and u … w, the corresponding reciprocal powers of the units for voltage (V) and temperature (K) must be used, so that the exponent expressions each yield a dimensionless result of "1":

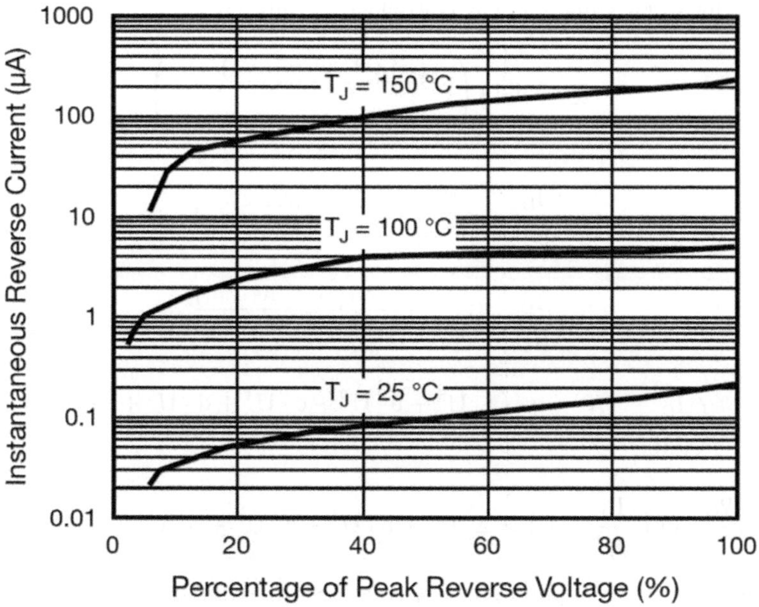

Fig. 8.16 Reverse characteristic field of a semiconductor diode [22]

$$A = A \cdot \left(e^{\frac{A \cdot s \cdot \left(\frac{1}{V^4} \cdot V^5 + \frac{1}{V^3} \cdot V^4 + \frac{1}{V^2} \cdot V^3 + \frac{1}{V} \cdot V^2 + 1 \cdot V \right)}{V \cdot A \cdot s}} - 1 \right) \cdot \left(e^{\frac{\frac{V \cdot A \cdot s}{K} \cdot \left(\frac{1}{K} \cdot K^2 + 1 \cdot K + K \right)}{V \cdot A \cdot s}} - 1 \right) \tag{8.9}$$

That is,

$$S_m \left(T_m / \frac{W_i}{k} \cdot \ln \left(\frac{I_D}{I_{D1}} + 1 \right) \right). \tag{8.10}$$

Unfortunately, the supporting points cannot simply be taken from the diagrams and inserted into the two systems of equations in Appendix 1. The following preliminary considerations and analyses are necessary to determine the supporting points for calculating the coefficients a … f and u … w:

For which temperature does the U-I diagram in the diode's datasheet apply (read from the datasheet)?

For this temperature, the following expression applies:

$$\left(e^{\frac{k \cdot \left(u \cdot T^2 + v \cdot T + w \right)}{W_i}} - 1 \right) = 1. \tag{8.11}$$

This allows the diode equation to be rearranged as follows:

$$I_D = |I_S| \cdot \left(e^{\frac{q \cdot \left(a \cdot U^5 + b \cdot U^4 + c \cdot U^3 + d \cdot U^2 + e \cdot U + f \right)}{W_i}} - 1 \right) \tag{8.12}$$

$$\frac{I_D}{|I_S|} = e^{\frac{q \cdot \left(a \cdot U^5 + b \cdot U^4 + c \cdot U^3 + d \cdot U^2 + e \cdot U + f \right)}{W_i}} - 1 \tag{8.13}$$

$$\frac{I_D}{|I_S|} + 1 = e^{\frac{q \cdot \left(a \cdot U^5 + b \cdot U^4 + c \cdot U^3 + d \cdot U^2 + e \cdot U + f \right)}{W_i}} \tag{8.14}$$

$$\ln\left(\frac{I_D}{|I_S|} + 1\right) = \frac{q \cdot \left(a \cdot U^5 + b \cdot U^4 + c \cdot U^3 + d \cdot U^2 + e \cdot U + f \right)}{W_i} \tag{8.15}$$

$$\frac{W_i}{q} \cdot \ln\left(\frac{I_D}{|I_S|} + 1\right) = a \cdot U^5 + b \cdot U^4 + c \cdot U^3 + d \cdot U^2 + e \cdot U + f \tag{8.16}$$

Accordingly, for the six support points used to determine the interpolation polynomial of the voltage, the following applies:

$$S_n\left(U_n / \frac{W_i}{q} \cdot \ln\left(\frac{I_{Dn}}{|I_S|} + 1\right)\right) \tag{8.17}$$

with $n = 1 \ldots 6$.

With these support points, the coefficients a … f can be determined in accordance with Chap. 9. From an application engineering perspective, it is usually not necessary for a support point to be determined for the voltage $U = 0$.

To determine the coefficients u … w, the expression

$$\left(e^{\frac{k \cdot \left(u \cdot T^2 + v \cdot T + w \right)}{W_i}} - 1 \right) = 1 \tag{8.18}$$

is used and rearranged (where T_1 is the temperature for which the graph of the U/I diagram applies):

$$\left(e^{\frac{k \cdot \left(u \cdot T_1^2 + v \cdot T_1 + w \right)}{W_i}} - 1 \right) = 1 = R_1(T_1) \tag{8.19}$$

$$\left(e^{\frac{k \cdot \left(u \cdot T_m^2 + v \cdot T_m + w \right)}{W_i}} - 1 \right) = R_m(T_m) \tag{8.20}$$

with $m \in [2, 3]$.

Furthermore, the following is calculated

$$\left(e^{\frac{k\cdot\left(u\cdot T_m^2+v\cdot T_m+w\right)}{W_i}}-1\right)=R_m(T_m) \tag{8.21}$$

$$e^{\frac{k\cdot\left(u\cdot T_m^2+v\cdot T_m+w\right)}{W_i}}=R_m(T_m)+1 \tag{8.22}$$

$$\frac{k\cdot\left(u\cdot T_m^2+v\cdot T_m+w\right)}{W_i}=\ln\left(R_m(T_m)+1\right) \tag{8.23}$$

$$u\cdot T_m^2+v\cdot T_m+w=\frac{W_i}{k}\cdot\ln\left(R_m(T_m)+1\right) \tag{8.24}$$

with

$$R_1(T_1)=1 \tag{8.25}$$

$$R_m(T_m)=\frac{I_{Dm}}{I_{D1}}\quad \text{with } m\ \varepsilon\ [2,3] \tag{8.26}$$

The values for the temperatures and currents are taken from the diagram at the same voltage.

The three support points for determining the coefficients u … w of the interpolation polynomial for the temperature are thus

$$S_m\left(T_m/\frac{W_i}{k}\cdot\ln\left(\frac{I_{Dm}}{I_{D1}}+1\right)\right) \tag{8.27}$$

with $m=1 \ldots 3$.

This means that the necessary support points for the voltage-current behavior and for the temperature can be taken from a diode's datasheet, and the coefficients for the interpolation polynomials can be determined using a prepared Excel spreadsheet (see also Sect. 8.3).

This diode equation only describes the behavior of a diode with sufficient accuracy within a certain range of voltage and current. The main reason for this is that any interpolation of a function using a polynomial can only achieve limited accuracy, not least because polynomials have minima and maxima, and the active range of the polynomial for interpolation must be chosen so that these minima and maxima do not interfere.

However, interpolation with another function, such as an exponential or trigonometric function, very quickly leads to transcendental systems of equations, which can only be solved by numerical iteration for fixed parameters. This, however, contradicts the intention of providing a general solution for modeling diode behavior, which can be applied within the normal operating range of the diode in the three-dimensional space defined by forward voltage, forward current, and temperature. This solution is applicable due to the correct assignment of the development of diode behavior in relation to forward voltage, forward current, and temperature, as opposed to the incorrect assignment in the Shockley equation. Some examples are presented in Sect. 8.4.

If higher accuracy or a certain minimum accuracy is required over a wider voltage or current range than can be provided by the fifth-order polynomial in the diode equation, it may also be possible to use an interpolation polynomial of higher order. The basic algorithm does not change as a result, but it should be noted that the modeled diode behavior will differ from the real diode in that local maxima and minima, or at least inflection points, may appear between the support points in the model, because the graph of the modeled equation "oscillates" around the real graph.

In today's era of digital computers with extensive data storage and high computational power, it is no longer common practice in most cases to simulate diode behavior using an analytical equation. Thus, the development of an analytical diode equation may not seem as significant at first glance as it was in the days of Mr. Shockley. Nowadays, depending on the desired level of accuracy, a certain number of reference points are taken from the diode characteristic diagrams and processed by the microprocessor during simulation. Intermediate values are usually determined by linear interpolation between the two nearest reference points.

However, to technically analyze and understand the behavior of a semiconductor diode, I still consider the analytical approach to be both meaningful and necessary.

8.3 Calculation of the Diode Equation

To represent the diode equation and compare it with data from real diode diagrams, an Excel spreadsheet is structured accordingly (Fig. 8.17).

As described in Sect. 8.2, the interpolation polynomials used to approximate the diode equation are not applied directly, but rather in the exponent of the exponential function. Accordingly, the coefficients for the interpolation polynomials are not taken directly from the reference points read from the diagrams, but are determined using a logarithmic function (for details, see Sect. 8.2).

The reference points are taken from the diagram for the corresponding diode, e.g., 1N4148, and entered into the Excel spreadsheet.

Corresponding formulas for calculation in Excel are derived from the formulas for the interpolation polynomials. To reduce the error rate, the procedure is as follows:

1. The expressions for the coefficients of the interpolation polynomials are created using a text editor; the coefficient expressions are transferred into Excel syntax using the existing variable representations (X1, X2 ... Y1, Y2 ... etc.), for example:
   K0 = -1*(X2*X3*X4*X5*X6*Y1)/((X1-X2)*(X1-X3)*(X1-X4)*(X1-X5)*(X1-X6))-(X1*X3*X4*X5*X6*Y2)/((X2-X1)*(X2-

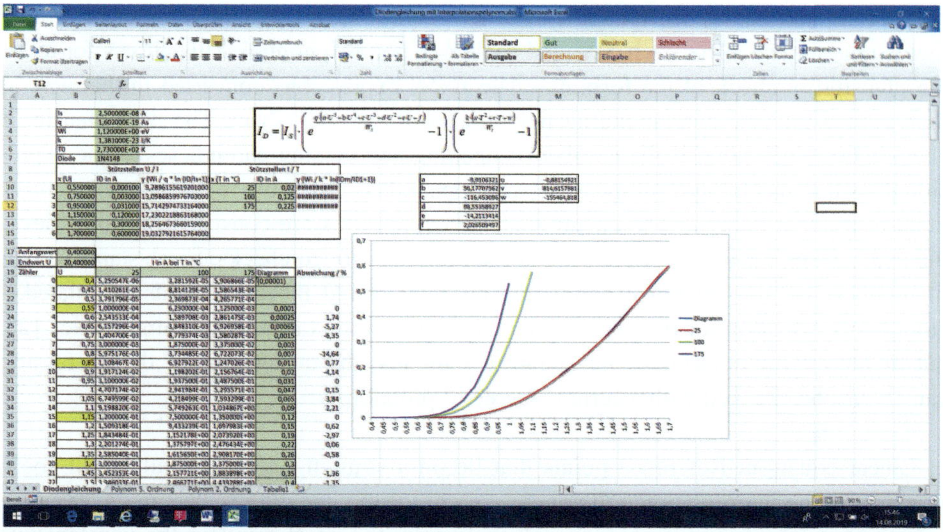

Fig. 8.17 Screenshot of the Excel spreadsheet for representing the diode equation and comparing diode data; green fields are input fields; in the upper left, the diode's key parameters from the datasheet are entered, to the right, for reference, the diode equation; below, in two sections, the reference points from the datasheet diagrams are entered; in the large field below, the top row is for entering temperatures, and in the column to the right below, the values read from the datasheet characteristic curve; at the bottom right, the characteristic curve field of the diode under consideration appears; above this, the parameters for the two polynomials are listed

X3)*(X2-X4)*(X2-X5)*(X2-X6))-(X1*X2*X4*X5*X6*Y3)/((X3-X1)*(X3-X2)*(X3-X4)*(X3-X5)*(X3-X6))-(X1*X2*X3*X5*X6*Y4)/((X4-X1)*(X4-X2)*(X4-X3)*(X4-X5)*(X4-X6))-(X1*X2*X3*X4*X6*Y5)/((X5-X1)*(X5-X2)*(X5-X3)*(X5-X4)*(X5-X6))-(X1*X2*X3*X4*X5*Y6)/((X6-X1)*(X6-X2)*(X6-X3)*(X6-X4)*(X6-X5)).

2. Next, using the "Find and Replace" function of the text editor, the variable names X1, X2, etc. are converted into the Excel cell references where the values for X1, X2, etc. are stored, e.g., B10, B11, etc.:
   K0 = -1*(B11*B12*B13*B14*B15*D10)/((B10-B11)*(B10-B12)*(B10-B13)*(B10-B14)*(B10-B15))-(B10*B12*B13*B14*B15*D11)/((B11-B10)*(B11-B12)*(B11-B13)*(B11-B14)*(B11-B15))-(B10*B11*B13*B14*B15*D12)/((B12-B10)*(B12-B11)*(B12-B13)*(B12-B14)*(B12-B15))-(B10*B11*B12*B14*B15*D13)/((B13-B10)*(B13-B11)*(B13-B12)*(B13-B14)*(B13-B15))-(B10*B11*B12*B13*B15*D14)/((B14-B10)*(B14-B11)*(B14-B12)*(B14-B13)*(B14-B15))-(B10*B11*B12*B13*B14*D15)/((B15-B10)*(B15-B11)*(B15-B12)*(B15-B13)*(B15-B14)).

If, for this procedure, all coefficient expressions are selected at once in the text editor and "Replace All" is used, then for all coefficients (K0 … K5 for the 5th order polynomial and K0 … K2 for the 2nd order polynomial), the same number of replacements must be made for each X… and each Y… per polynomial. This allows for a simple plausibility check of the formulas. The Excel expressions converted in this way are then copied into the corresponding fields in the Excel worksheet. The next plausibility check takes place here: if the number of opening and closing parentheses in the expression does not match, an error message will appear. It is advisable to always make corrections in the text document to ensure that the expressions are consistent in both files. If the expressions for the coefficients pass the aforementioned checks, some standard pairs of numbers for the reference points can be entered and the coefficients calculated. A complete verification of the expressions through testing is not possible in practice [23], but by selectively entering certain pairs of numbers, the correctness of the expressions can at least be checked with a certain degree of probability:

1. Linear equation: Reference points that lie on a straight line will result in all coefficients of the interpolation polynomial, except for the linear term, becoming "0"; if the reference line does not pass through the origin, the coefficient for the constant term will also be nonzero.
2. The x-values of some sets of reference points are substituted into the polynomial; the corresponding y-values of the reference points must appear as the result.

In addition to the Excel worksheet for calculating and displaying the diagram for a diode, there is a worksheet in which the 5th order polynomial is plotted within a specific range (Fig. 8.18). This makes it easy to see whether the polynomial has a local extremum and/or an inflection point in the value range relevant to the diode equation.

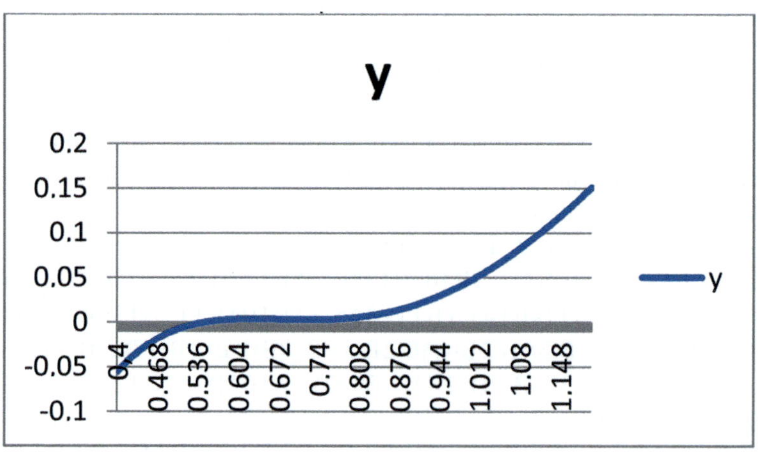

Fig. 8.18 Example graph in Excel for a 5th order polynomial, to determine whether there are local extrema in the relevant range of the function

8.4 Analysis of Various Diode Examples

The examples of diode characteristic curves presented in this appendix are taken from Vishay datasheets. This is primarily because the Vishay datasheets provide the information and diagrams necessary for modeling, at least for the diodes presented here.

The procedure for determining the coefficients for the diode equation in the following examples is always the same: From the diagram or family of curves showing the forward current as a function of the forward voltage—almost always with crystal temperature as a parameter—the 25 °C characteristic is transferred into a table in 0.05 V increments. From these pairs of voltage and current values, six are selected and entered into the table presented in Sect. 8.3. At the same time, the values for voltages and currents are entered into a table. In addition, supporting points for the thermal behavior of the current are taken from the diagram and entered into the table presented in Sect. 8.3.

In the resulting diagram, both the recorded and the calculated value pairs are shown. Furthermore, for the additional temperatures—besides the 25 °C characteristic—the characteristic curves are calculated and displayed in the diagram.

The model characteristic curves are shown for the diodes listed in Table 8.1.

Tab. 8.1 List of the diodes modeled below [24–37]

Diode Name	Function Type	Saturation Reverse Current/A	Figures
1N4148	Small Signal Fast Switching Diode	$2.5 \cdot 10^{-8}$	8.19
AU1PM	Ultrafast Avalanche Rectifier	$1.0 \cdot 10^{-6}$	8.20
BAT42W	Small Signal Schottky Diode	$5.0 \cdot 10^{-7}$	8.21
BAT46W	Small Signal Schottky Diode	$8.0 \cdot 10^{-7}$	8.22
BU1006	Bridge Rectifier	$5.0 \cdot 10^{-6}$	8.23
BYV98	Ultrafast Avalanche Diode	$1.0 \cdot 10^{-5}$	8.24
LL46	Small Signal Schottky Diode	$8.0 \cdot 10^{-7}$	8.25
MBR10H	High Voltage Schottky Diode	$4.5 \cdot 10^{-6}$	8.26
PB5006	Bridge Rectifier	$1.0 \cdot 10^{-5}$	8.27
SF1200	Ultrafast Avalanche Diode	$5.0 \cdot 10^{-6}$	8.28
UHF20-FCT	Ultrafast Recovery Rectifier	$5.0 \cdot 10^{-6}$	8.29
V20202C	High Voltage Trench MOS Barrier Schottky Rectifier	$4.0 \cdot 10^{-5}$	8.30
VS-150EBU02HF4	Ultrafast Soft Recovery Diode	$5.0 \cdot 10^{-5}$	8.31
VS-SD853	Fast Recovery Diode, Puck Version	$1.0 \cdot 10^{-1}$	8.32

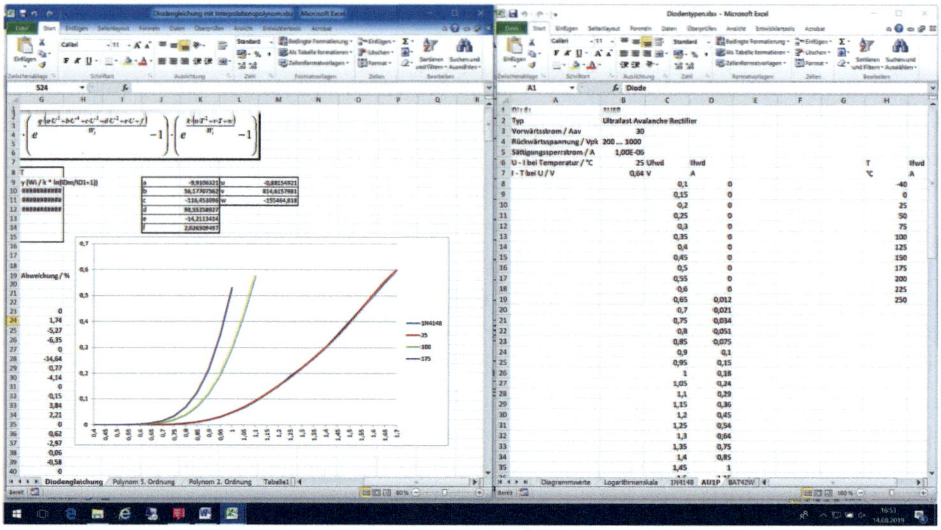

Fig. 8.19 Family of curves for a 1N4148; values taken from a datasheet (blue = diagram) are shown together with calculated values for 25 °C (red), 100 °C (green), and 175 °C (violet)

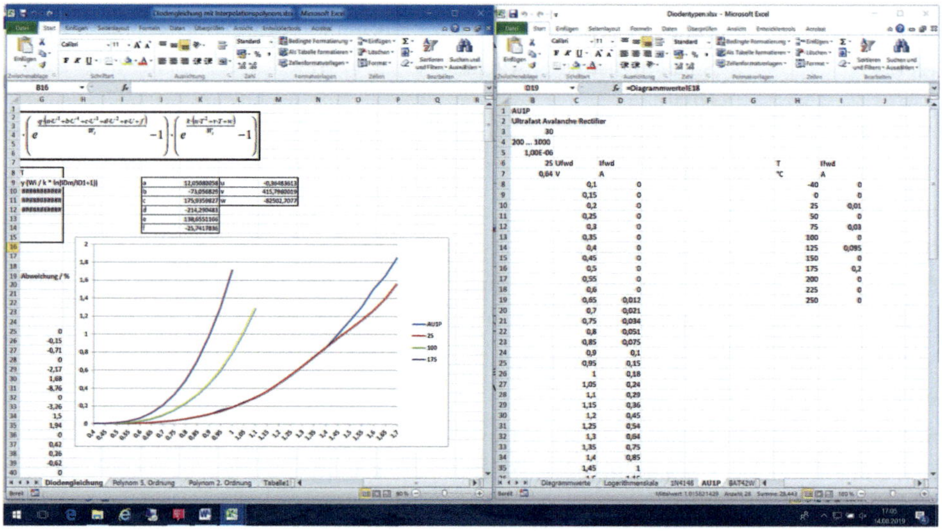

Fig. 8.20 Family of curves for an AU1P; values taken from a datasheet (blue = diagram) are shown together with calculated values for 25 °C (red), 100 °C (green), and 175 °C (violet)

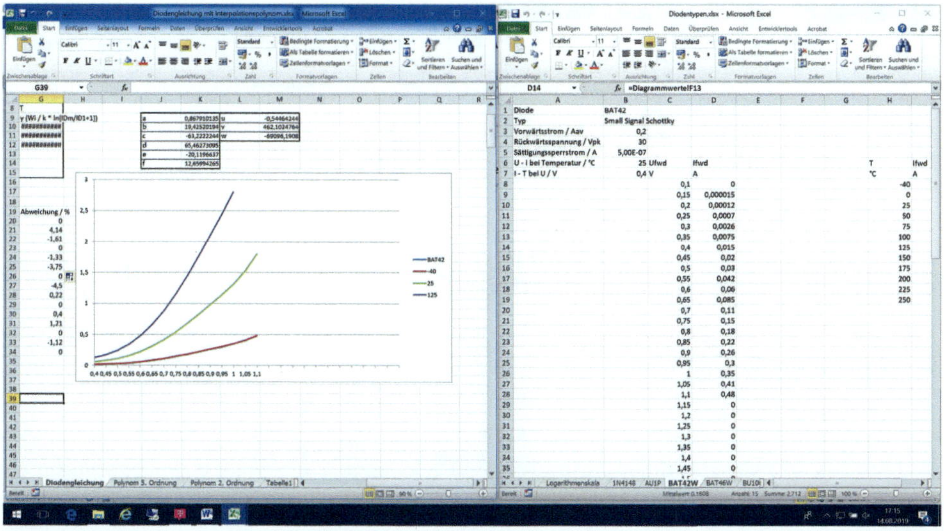

Fig. 8.21 Family of curves for a BAT42; values taken from a datasheet (blue = diagram) are shown together with calculated values for 25 °C (red), 100 °C (green), and 175 °C (violet)

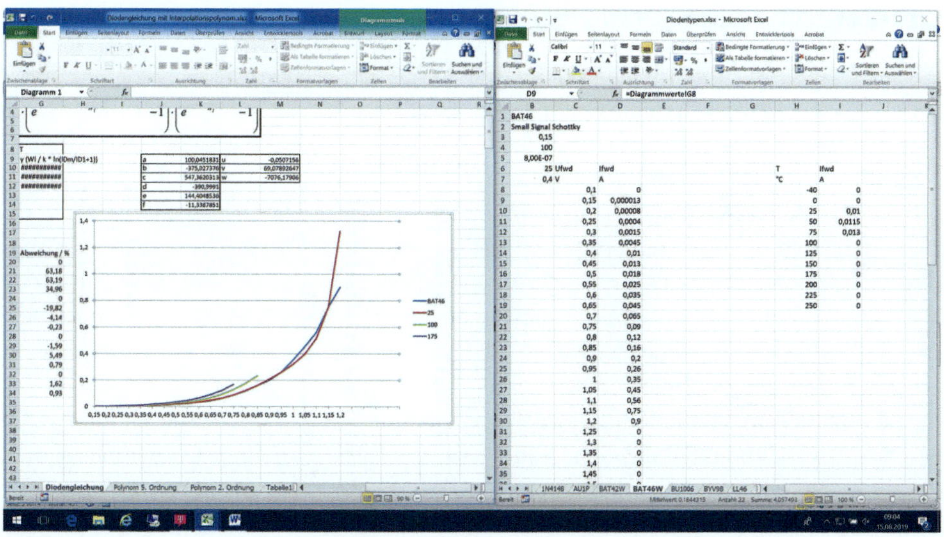

Fig. 8.22 Family of curves for a BAT46; values taken from a datasheet (blue = diagram) are shown together with calculated values for 25 °C (red), 100 °C (green), and 175 °C (violet)

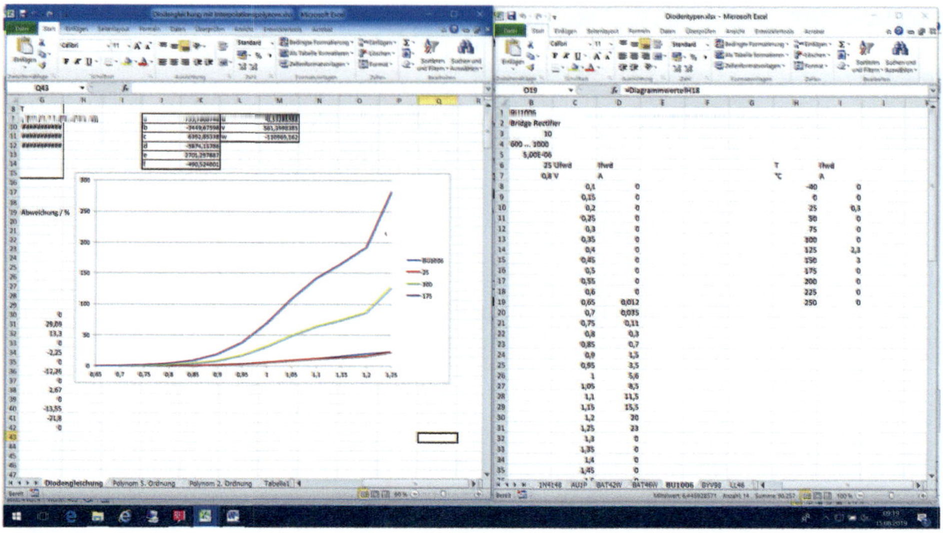

Fig. 8.23 Family of curves for a BU1006; values taken from a datasheet (blue = diagram) are shown together with calculated values for 25 °C (red), 100 °C (green), and 175 °C (violet)

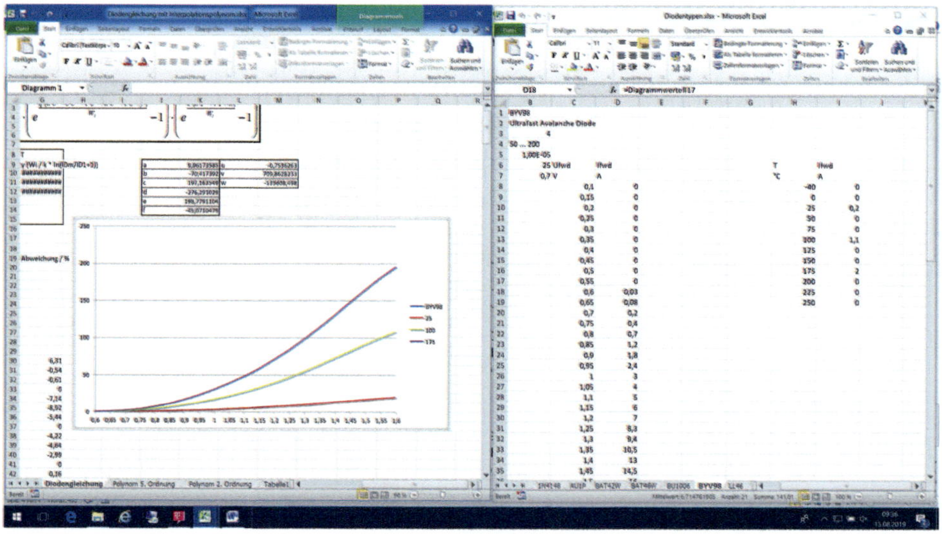

Fig. 8.24 Family of curves for a BYV98; values taken from a datasheet (blue = diagram) are shown together with calculated values for 25 °C (red), 100 °C (green), and 175 °C (violet)

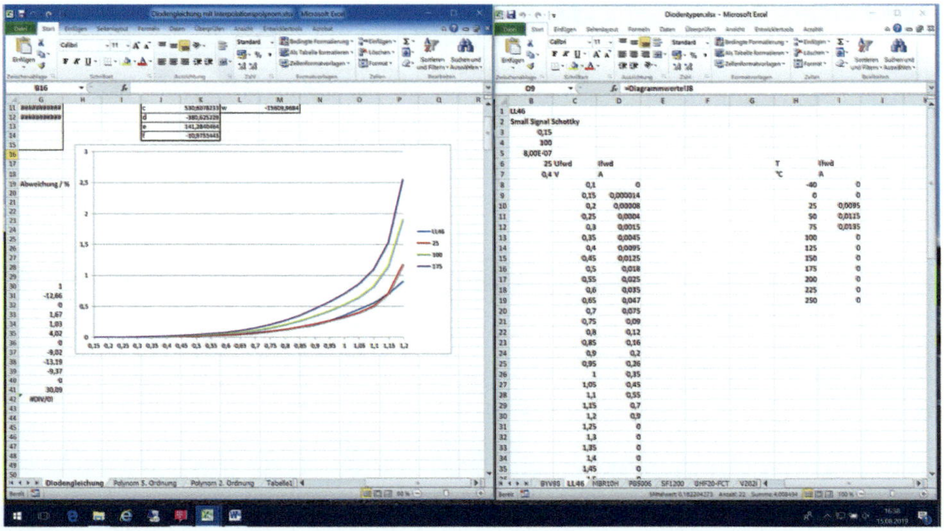

Fig. 8.25 Family of curves for an LL46; values taken from a datasheet (blue = diagram) are shown together with calculated values for 25 °C (red), 100 °C (green), and 175 °C (violet)

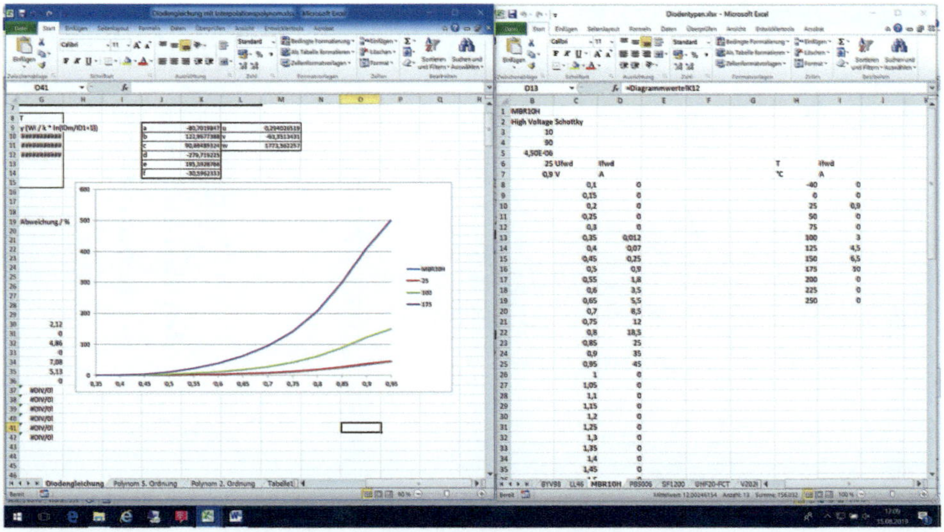

Fig. 8.26 Family of curves for an MBR10H; values taken from a datasheet (blue = diagram) are shown together with calculated values for 25 °C (red), 100 °C (green), and 175 °C (violet)

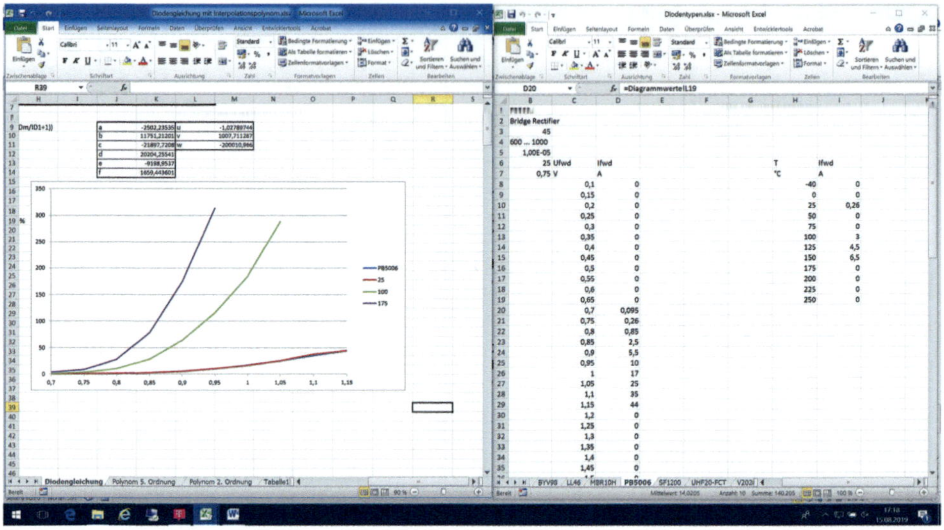

Fig. 8.27 Family of curves for a PB5006; values taken from a datasheet (blue = diagram) are compared with calculated values for 25 °C (red), 100 °C (green), and 175 °C (violet)

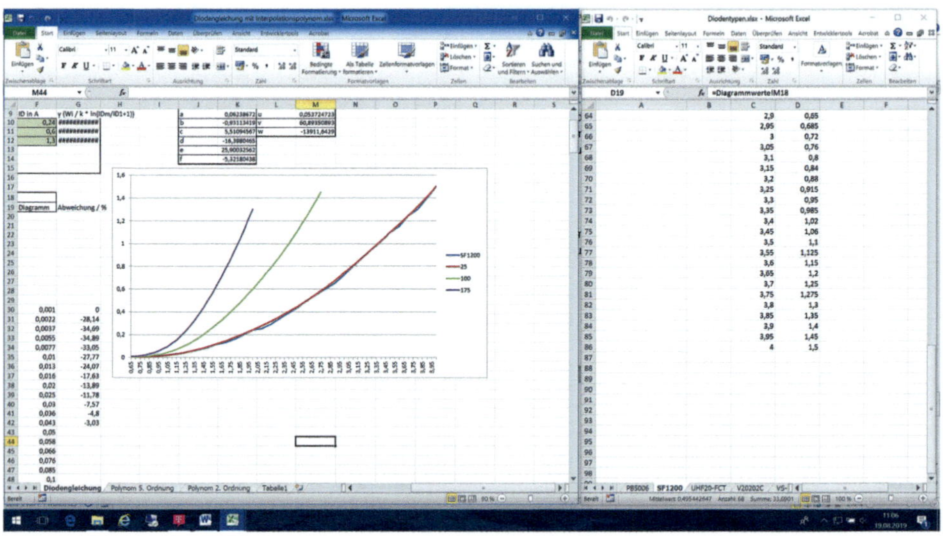

Fig. 8.28 Family of curves for an SF1200; values taken from a datasheet (blue = diagram) are compared with calculated values for 25 °C (red), 100 °C (green), and 175 °C (violet)

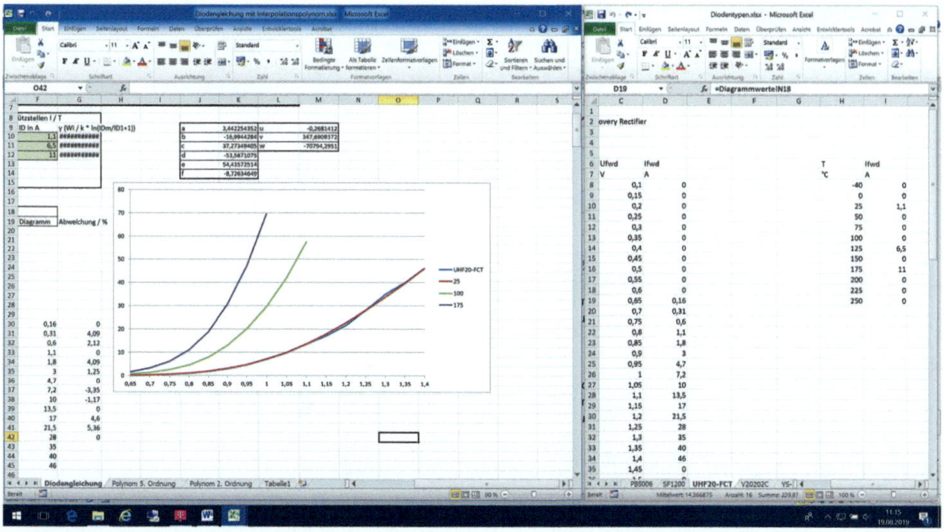

Fig. 8.29 Family of curves for a UHF20-FCT; values taken from a datasheet (blue = diagram) are compared with calculated values for 25 °C (red), 100 °C (green), and 175 °C (violet)

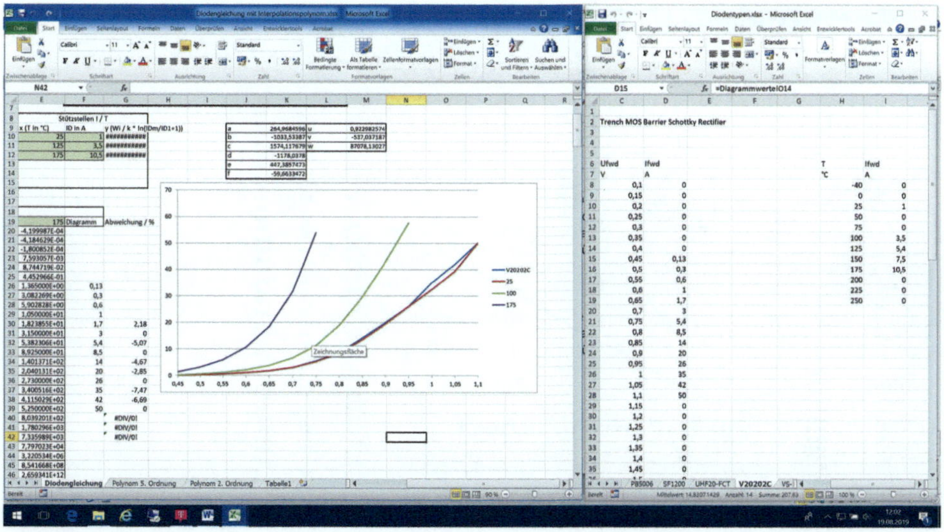

Fig. 8.30 Family of curves for a V20202C; values taken from a datasheet (blue = diagram) are compared with calculated values for 25 °C (red), 100 °C (green), and 175 °C (violet)

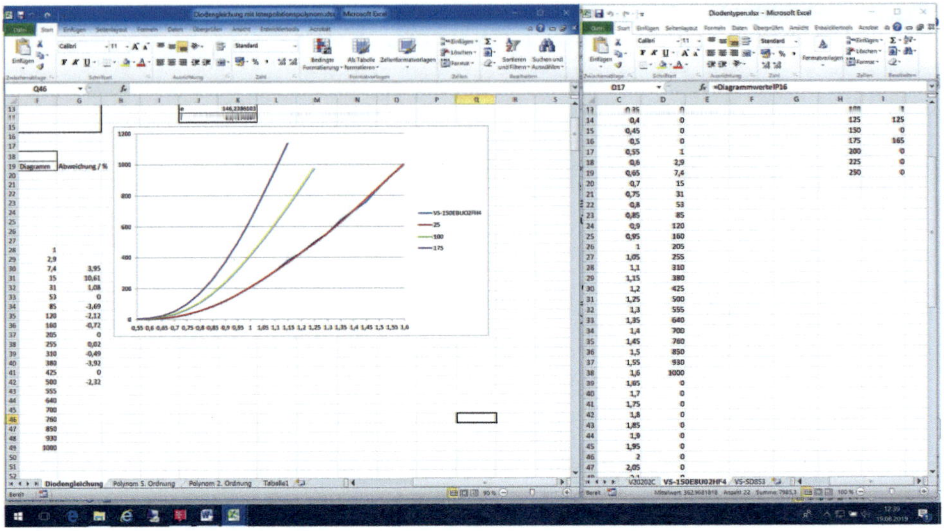

Fig. 8.31 Family of curves for a VS-150EBUO2FH4; values taken from a datasheet (blue = diagram) are compared with calculated values for 25 °C (red), 100 °C (green), and 175 °C (violet)

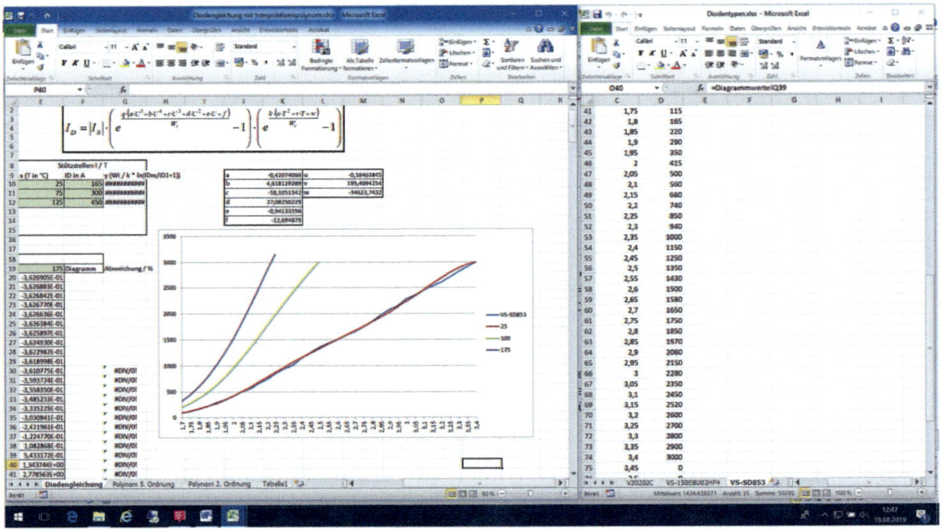

Fig. 8.32 Family of curves for a VS-SD853; values taken from a datasheet (blue = diagram) are compared with calculated values for 25 °C (red), 100 °C (green), and 175 °C (violet)

References

1. Fernuniversität Hagen, Studienskript „Halbleiter Schaltungstechnik", 1990
2. Ulrick Tietze, Christoph Schenk et al; Halbleiter-Schaltungstechnik; Springer Verlag 1993
3. Wikipedia, Edison-Richardson-Effekt, 1. 3. 2021
4. Wikipedia, Feldemission, 1. 3. 2021
5. Josef von Stackelberg et al; Handbuch Elektrofilter; Springer Verlag, 2018
6. R. Müller, Grundlagen der Halbleiter-Elektronik, Springer-Verlag, 1991
7. Abbildung 11 in R. Müller, Grundlagen der Halbleiter-Elektronik, Springer-Verlag, 1991
8. Abbildung 48 in R. Müller, Grundlagen der Halbleiter-Elektronik, Springer-Verlag, 1991
9. Abbildung 15 in R. Müller, Grundlagen der Halbleiter-Elektronik, Springer-Verlag, 1991
10. Abbildung 24 in R. Müller, Grundlagen der Halbleiter-Elektronik, Springer-Verlag, 1991
11. Abbildung 16 in R. Müller, Grundlagen der Halbleiter-Elektronik, Springer-Verlag, 1991
12. Abbildung 20 in R. Müller, Grundlagen der Halbleiter-Elektronik, Springer-Verlag, 1991
13. Abbildung 21 in R. Müller, Grundlagen der Halbleiter-Elektronik, Springer-Verlag, 1991
14. Abbildung 23 in R. Müller, Grundlagen der Halbleiter-Elektronik, Springer-Verlag, 1991
15. Abbildung 9 in Vishay, Data Sheet VS-SD853C ..S50K Series
16. Abbildung 25 in R. Müller, Grundlagen der Halbleiter-Elektronik, Springer-Verlag, 1991
17. Abbildung 77 in R. Müller, Grundlagen der Halbleiter-Elektronik, Springer-Verlag, 1991
18. Wikipedia, Shockley-Gleichung, 1. 3. 2021
19. ON Semi, Datenblatt der 1N4001, Figur 1
20. Vishay, Datenblatt der 1N4148, Figur 2
21. Vishay, Datenblatt der 1N4148, Figur 1
22. Vishay, Datenblatt der 1N4001, Figur 5
23. Josef von Stackelberg, Hardware-Interpretation graphisch formulierter sicherheitsgerichteter Echtzeitprogramme, VDI-Verlag, 2006
24. Vishay, Datenblatt der 1N4148, Version von 12/2013
25. Vishay, Datenblatt der AU1PM, Version von 02/2015
26. Vishay, Datenblatt der BAT42W, Version von 02/2013
27. Vishay, Datenblatt der BAT46W, Version von 02/2013
28. Vishay, Datenblatt der BU1006, Version von 08/2013
29. Vishay, Datenblatt der BYV98, Version von 09/2012
30. Vishay, Datenblatt der LL46, Version von 05/2012
31. Vishay, Datenblatt der MBR10H, Version von 07/2015
32. Vishay, Datenblatt der PB5006, Version von 06/2013
33. Vishay, Datenblatt der SF1200, Version von 09/2012
34. Vishay, Datenbl. der UHF20-FCT, Version von 08/2013
35. Vishay, Datenblatt der V20202C, Version von 10/2014
36. Vishay, Dat.-bl. der VS-150EBU02HF4, Vers. v. 06/2015
37. Vishay, Datenblatt der VS-SD853, Version von 04/2014

9 The Interpolation Polynomial

Functional relationships in nature are rarely such that they can be mathematically modeled with satisfactory accuracy over a wide range of a given dimension, primarily because the mathematical functions quickly become unwieldy. In order to still be able to represent the functional relationships of nature mathematically, interpolation methods can be employed. In these methods, a mathematical function is determined based on a number of specific points within the value range under consideration, which at least at these specific points represents the functional relationship with sufficient accuracy. Naturally, in the regions between these specific points, the mathematical function thus determined will deviate from nature, as is the very nature of a model. The more of these specific points, called nodes, are used to determine the mathematical function, the greater the chance of reducing the difference between the mathematical model and natural reality. However, this also means that arbitrarily increasing the number of nodes does not necessarily imply that the mathematical model will exhibit a smaller deviation from reality at all points between the nodes than with fewer nodes. This also depends on the choice of the model function used for this so-called "interpolation function."

In principle, any arbitrary function can be selected, equipped with the necessary parameters, the nodes substituted, and the calculation algorithm applied. Sooner or later, however, this will lead to transcendental equations, which can only be solved approximately using iterative methods, and thus do not provide a satisfactory solution for the parameters.

For this reason, polynomials are used in most cases for interpolation tasks. Although these too, once they reach higher order, tend to become unwieldy, they at least remain solvable. Apart from that, there are various solution schemes, such as those by Lagrange or Newton [1]. Other mathematicians, such as Gauss, offer solution schemes that allow the parameters of the solution polynomial to be determined from the nodes [2].

What I find lacking in all these solution schemes, however, is that they do not explicitly represent, for a given coefficient of the solution polynomial, its dependence on the

J. von Stackelberg, *The Mass of a Photon*, https://doi.org/10.1007/978-3-658-49408-7_9

coordinate values of the nodes. For this reason, I attempted to determine this dependence myself, first for the fifth-order polynomial and then for the second-order polynomial, as I needed these in the initial approach for the diode equation (for details, see Chap. 8). The solutions for the coefficients of the two polynomials were as follows:

Second-order polynomial with three nodes:

$$y = x^2 \cdot a + x \cdot b + c \tag{9.1}$$

$$S_1(x_1/y_1)$$

$$S_2(x_2/y_2)$$

$$S_3(x_3/y_3)$$

$$a = \frac{y_1}{(x_1 - x_2) \cdot (x_1 - x_3)} + \frac{y_2}{(x_2 - x_1) \cdot (x_2 - x_3)} + \frac{y_3}{(x_3 - x_1) \cdot (x_3 - x_2)} \tag{9.2}$$

$$b = -\frac{y_1 \cdot (x_2 + x_3)}{(x_1 - x_2) \cdot (x_1 - x_3)} - \frac{y_2 \cdot (x_1 + x_3)}{(x_2 - x_1) \cdot (x_2 - x_3)} - \frac{y_3 \cdot (x_1 + x_2)}{(x_3 - x_1) \cdot (x_3 - x_2)} \tag{9.3}$$

$$c = \frac{y_1 \cdot (x_2 \cdot x_3)}{(x_1 - x_2) \cdot (x_1 - x_3)} + \frac{y_2 \cdot (x_1 \cdot x_3)}{(x_2 - x_1) \cdot (x_2 - x_3)} + \frac{y_3 \cdot (x_1 \cdot x_2)}{(x_3 - x_1) \cdot (x_3 - x_2)} \tag{9.4}$$

The fifth-order polynomial with six support points:

$$y = x^5 \cdot a + x^4 \cdot b + x^3 \cdot c + x^2 \cdot d + x \cdot e + f \tag{9.5}$$

$$S_1(x_1/y_1)$$

$$S_2(x_2/y_2)$$

$$S_3(x_3/y_3)$$

$$S_4(x_4/y_4)$$

$$S_5(x_5/y_5)$$

$$S_6(x_6/y_6)$$

$$
\begin{aligned}
a = & \frac{y_1}{(x_1 - x_2) \cdot (x_1 - x_3) \cdot (x_1 - x_4) \cdot (x_1 - x_5) \cdot (x_1 - x_6)} + \\
& + \frac{y_2}{(x_2 - x_1) \cdot (x_2 - x_3) \cdot (x_2 - x_4) \cdot (x_2 - x_5) \cdot (x_2 - x_6)} + \\
& + \frac{y_3}{(x_3 - x_1) \cdot (x_3 - x_2) \cdot (x_3 - x_4) \cdot (x_3 - x_5) \cdot (x_3 - x_6)} + \\
& + \frac{y_4}{(x_4 - x_1) \cdot (x_4 - x_2) \cdot (x_4 - x_3) \cdot (x_4 - x_5) \cdot (x_4 - x_6)} + \\
& + \frac{y_5}{(x_5 - x_1) \cdot (x_5 - x_2) \cdot (x_5 - x_3) \cdot (x_5 - x_4) \cdot (x_5 - x_6)} + \\
& + \frac{y_6}{(x_6 - x_1) \cdot (x_6 - x_2) \cdot (x_6 - x_3) \cdot (x_6 - x_4) \cdot (x_6 - x_5)}
\end{aligned}
\tag{9.6}
$$

$$
\begin{aligned}
b = & -\frac{y_1 \cdot (x_6 + x_5 + x_4 + x_3 + x_2)}{(x_1 - x_2) \cdot (x_1 - x_3) \cdot (x_1 - x_4) \cdot (x_1 - x_5) \cdot (x_1 - x_6)} - \\
& - \frac{y_2 \cdot (x_6 + x_5 + x_4 + x_3 + x_1)}{(x_2 - x_1) \cdot (x_2 - x_3) \cdot (x_2 - x_4) \cdot (x_2 - x_5) \cdot (x_2 - x_6)} - \\
& - \frac{y_3 \cdot (x_6 + x_5 + x_4 + x_2 + x_1)}{(x_3 - x_1) \cdot (x_3 - x_2) \cdot (x_3 - x_4) \cdot (x_3 - x_5) \cdot (x_3 - x_6)} - \\
& - \frac{y_4 \cdot (x_6 + x_5 + x_3 + x_2 + x_1)}{(x_4 - x_1) \cdot (x_4 - x_2) \cdot (x_4 - x_3) \cdot (x_4 - x_5) \cdot (x_4 - x_6)} - \\
& - \frac{y_5 \cdot (x_6 + x_4 + x_3 + x_2 + x_1)}{(x_5 - x_1) \cdot (x_5 - x_2) \cdot (x_5 - x_3) \cdot (x_5 - x_4) \cdot (x_5 - x_6)} - \\
& - \frac{y_6 \cdot (x_5 + x_4 + x_3 + x_2 + x_1)}{(x_6 - x_1) \cdot (x_6 - x_2) \cdot (x_6 - x_3) \cdot (x_6 - x_4) \cdot (x_6 - x_5)}
\end{aligned}
\tag{9.7}
$$

$$
\begin{aligned}
c = & \frac{y_1 \cdot (x_2 \cdot x_3 + x_2 \cdot x_4 + x_2 \cdot x_5 + x_2 \cdot x_6 + x_3 \cdot x_4 + x_3 \cdot x_5 + x_3 \cdot x_6 + x_4 \cdot x_5 + x_4 \cdot x_6 + x_5 \cdot x_6)}{(x_1 - x_2) \cdot (x_1 - x_3) \cdot (x_1 - x_4) \cdot (x_1 - x_5) \cdot (x_1 - x_6)} + \\
& + \frac{y_2 \cdot (x_1 \cdot x_3 + x_1 \cdot x_4 + x_1 \cdot x_5 + x_1 \cdot x_6 + x_3 \cdot x_4 + x_3 \cdot x_5 + x_3 \cdot x_6 + x_4 \cdot x_5 + x_4 \cdot x_6 + x_5 \cdot x_6)}{(x_2 - x_1) \cdot (x_2 - x_3) \cdot (x_2 - x_4) \cdot (x_2 - x_5) \cdot (x_2 - x_6)} + \\
& + \frac{y_3 \cdot (x_1 \cdot x_2 + x_1 \cdot x_4 + x_1 \cdot x_5 + x_1 \cdot x_6 + x_2 \cdot x_4 + x_2 \cdot x_5 + x_2 \cdot x_6 + x_4 \cdot x_5 + x_4 \cdot x_6 + x_5 \cdot x_6)}{(x_3 - x_1) \cdot (x_3 - x_2) \cdot (x_3 - x_4) \cdot (x_3 - x_5) \cdot (x_3 - x_6)} + \\
& + \frac{y_4 \cdot (x_1 \cdot x_2 + x_1 \cdot x_3 + x_1 \cdot x_5 + x_1 \cdot x_6 + x_2 \cdot x_3 + x_2 \cdot x_5 + x_2 \cdot x_6 + x_3 \cdot x_5 + x_3 \cdot x_6 + x_5 \cdot x_6)}{(x_4 - x_1) \cdot (x_4 - x_2) \cdot (x_4 - x_3) \cdot (x_4 - x_5) \cdot (x_4 - x_6)} + \\
& + \frac{y_5 \cdot (x_1 \cdot x_2 + x_1 \cdot x_3 + x_1 \cdot x_4 + x_1 \cdot x_6 + x_2 \cdot x_3 + x_2 \cdot x_4 + x_2 \cdot x_6 + x_3 \cdot x_4 + x_3 \cdot x_6 + x_4 \cdot x_6)}{(x_5 - x_1) \cdot (x_5 - x_2) \cdot (x_5 - x_3) \cdot (x_5 - x_4) \cdot (x_5 - x_6)} + \\
& + \frac{y_6 \cdot (x_1 \cdot x_2 + x_1 \cdot x_3 + x_1 \cdot x_4 + x_1 \cdot x_5 + x_2 \cdot x_3 + x_2 \cdot x_4 + x_2 \cdot x_5 + x_3 \cdot x_4 + x_3 \cdot x_5 + x_4 \cdot x_5)}{(x_6 - x_1) \cdot (x_6 - x_2) \cdot (x_6 - x_3) \cdot (x_6 - x_4) \cdot (x_6 - x_5)}
\end{aligned}
\tag{9.8}
$$

$$
\begin{aligned}
d = &-\frac{y_1 \cdot (x_2 \cdot x_3 \cdot x_4 + x_2 \cdot x_3 \cdot x_5 + x_2 \cdot x_3 \cdot x_6 + x_3 \cdot x_4 \cdot x_5 + x_3 \cdot x_4 \cdot x_6 + x_4 \cdot x_5 \cdot x_6)}{(x_1 - x_2) \cdot (x_1 - x_3) \cdot (x_1 - x_4) \cdot (x_1 - x_5) \cdot (x_1 - x_6)} - \\
&-\frac{y_2 \cdot (x_1 \cdot x_3 \cdot x_4 + x_1 \cdot x_3 \cdot x_5 + x_1 \cdot x_3 \cdot x_6 + x_3 \cdot x_4 \cdot x_5 + x_3 \cdot x_4 \cdot x_6 + x_4 \cdot x_5 \cdot x_6)}{(x_2 - x_1) \cdot (x_2 - x_3) \cdot (x_2 - x_4) \cdot (x_2 - x_5) \cdot (x_2 - x_6)} - \\
&-\frac{y_3 \cdot (x_1 \cdot x_2 \cdot x_4 + x_1 \cdot x_2 \cdot x_5 + x_1 \cdot x_2 \cdot x_6 + x_2 \cdot x_4 \cdot x_5 + x_2 \cdot x_4 \cdot x_6 + x_4 \cdot x_5 \cdot x_6)}{(x_3 - x_1) \cdot (x_3 - x_2) \cdot (x_3 - x_4) \cdot (x_3 - x_5) \cdot (x_3 - x_6)} - \\
&-\frac{y_4 \cdot (x_1 \cdot x_2 \cdot x_3 + x_1 \cdot x_2 \cdot x_5 + x_1 \cdot x_2 \cdot x_6 + x_2 \cdot x_3 \cdot x_5 + x_2 \cdot x_3 \cdot x_6 + x_3 \cdot x_5 \cdot x_6)}{(x_4 - x_1) \cdot (x_4 - x_2) \cdot (x_4 - x_3) \cdot (x_4 - x_5) \cdot (x_4 - x_6)} - \\
&-\frac{y_5 \cdot (x_1 \cdot x_2 \cdot x_3 + x_1 \cdot x_2 \cdot x_4 + x_1 \cdot x_2 \cdot x_6 + x_2 \cdot x_3 \cdot x_4 + x_2 \cdot x_3 \cdot x_6 + x_3 \cdot x_4 \cdot x_6)}{(x_5 - x_1) \cdot (x_5 - x_2) \cdot (x_5 - x_3) \cdot (x_5 - x_4) \cdot (x_5 - x_6)} - \\
&-\frac{y_6 \cdot (x_1 \cdot x_2 \cdot x_3 + x_1 \cdot x_2 \cdot x_4 + x_1 \cdot x_2 \cdot x_5 + x_2 \cdot x_3 \cdot x_4 + x_2 \cdot x_3 \cdot x_5 + x_3 \cdot x_4 \cdot x_5)}{(x_6 - x_1) \cdot (x_6 - x_2) \cdot (x_6 - x_3) \cdot (x_6 - x_4) \cdot (x_6 - x_5)}
\end{aligned}
\tag{9.9}
$$

$$
\begin{aligned}
e = &\frac{y_1 \cdot (x_2 \cdot x_3 \cdot x_4 \cdot x_5 + x_2 \cdot x_3 \cdot x_4 \cdot x_6 + x_3 \cdot x_4 \cdot x_5 \cdot x_6)}{(x_1 - x_2) \cdot (x_1 - x_3) \cdot (x_1 - x_4) \cdot (x_1 - x_5) \cdot (x_1 - x_6)} + \\
&+\frac{y_2 \cdot (x_1 \cdot x_3 \cdot x_4 \cdot x_5 + x_1 \cdot x_3 \cdot x_4 \cdot x_6 + x_3 \cdot x_4 \cdot x_5 \cdot x_6)}{(x_2 - x_1) \cdot (x_2 - x_3) \cdot (x_2 - x_4) \cdot (x_2 - x_5) \cdot (x_2 - x_6)} + \\
&+\frac{y_3 \cdot (x_1 \cdot x_2 \cdot x_4 \cdot x_5 + x_1 \cdot x_2 \cdot x_4 \cdot x_6 + x_2 \cdot x_4 \cdot x_5 \cdot x_6)}{(x_3 - x_1) \cdot (x_3 - x_2) \cdot (x_3 - x_4) \cdot (x_3 - x_5) \cdot (x_3 - x_6)} + \\
&+\frac{y_4 \cdot (x_1 \cdot x_2 \cdot x_3 \cdot x_5 + x_1 \cdot x_2 \cdot x_3 \cdot x_6 + x_2 \cdot x_3 \cdot x_5 \cdot x_6)}{(x_4 - x_1) \cdot (x_4 - x_2) \cdot (x_4 - x_3) \cdot (x_4 - x_5) \cdot (x_4 - x_6)} + \\
&+\frac{y_5 \cdot (x_1 \cdot x_2 \cdot x_3 \cdot x_4 + x_1 \cdot x_2 \cdot x_3 \cdot x_6 + x_2 \cdot x_3 \cdot x_4 \cdot x_6)}{(x_5 - x_1) \cdot (x_5 - x_2) \cdot (x_5 - x_3) \cdot (x_5 - x_4) \cdot (x_5 - x_6)} + \\
&+\frac{y_6 \cdot (x_1 \cdot x_2 \cdot x_3 \cdot x_4 + x_1 \cdot x_2 \cdot x_3 \cdot x_5 + x_2 \cdot x_3 \cdot x_4 \cdot x_5)}{(x_6 - x_1) \cdot (x_6 - x_2) \cdot (x_6 - x_3) \cdot (x_6 - x_4) \cdot (x_6 - x_5)}
\end{aligned}
\tag{9.10}
$$

$$
\begin{aligned}
f = &-\frac{y_1 \cdot (x_2 \cdot x_3 \cdot x_4 \cdot x_5 \cdot x_6)}{(x_1 - x_2) \cdot (x_1 - x_3) \cdot (x_1 - x_4) \cdot (x_1 - x_5) \cdot (x_1 - x_6)} - \\
&-\frac{y_2 \cdot (x_1 \cdot x_3 \cdot x_4 \cdot x_5 \cdot x_6)}{(x_2 - x_1) \cdot (x_2 - x_3) \cdot (x_2 - x_4) \cdot (x_2 - x_5) \cdot (x_2 - x_6)} - \\
&-\frac{y_3 \cdot (x_1 \cdot x_2 \cdot x_4 \cdot x_5 \cdot x_6)}{(x_3 - x_1) \cdot (x_3 - x_2) \cdot (x_3 - x_4) \cdot (x_3 - x_5) \cdot (x_3 - x_6)} - \\
&-\frac{y_4 \cdot (x_1 \cdot x_2 \cdot x_3 \cdot x_5 \cdot x_6)}{(x_4 - x_1) \cdot (x_4 - x_2) \cdot (x_4 - x_3) \cdot (x_4 - x_5) \cdot (x_4 - x_6)} - \\
&-\frac{y_5 \cdot (x_1 \cdot x_2 \cdot x_3 \cdot x_4 \cdot x_6)}{(x_5 - x_1) \cdot (x_5 - x_2) \cdot (x_5 - x_3) \cdot (x_5 - x_4) \cdot (x_5 - x_6)} - \\
&-\frac{y_6 \cdot (x_1 \cdot x_2 \cdot x_3 \cdot x_4 \cdot x_5)}{(x_6 - x_1) \cdot (x_6 - x_2) \cdot (x_6 - x_3) \cdot (x_6 - x_4) \cdot (x_6 - x_5)}
\end{aligned}
\tag{9.11}
$$

Since, upon examining the results, I suspected a general algorithm for determining the coefficients of the solution polynomial, I also calculated the sets of coefficients for polynomials of first, third, and fourth order. From these results, I derived the following general solution for the coefficients of an interpolation polynomial.

Interpolation polynomial of order n with n + 1 support points:

$$S(x_0/y_0)$$

$$S(x_1/y_1) \tag{9.12}$$

...

$$S(x_{n-1}/y_{n-1})$$

$$S(x_n/y_n)$$

$$y = k_n \cdot x^n + k_{n-1} \cdot x^{n-1} + \ldots + k_2 \cdot x^2 + k_1 \cdot x + k_0 \tag{9.13}$$

or

$$y = \sum_{m=0}^{n} k_m \cdot x^m \tag{9.14}$$

with

$$k_m = (-1)^n \cdot (-1)^m \cdot \sum_{j=0}^{n} \frac{y_j \cdot \sum_{k=1}^{\binom{n}{m}} \prod_{n=1}^{n-m} x_{p_1}}{(x_j - x_0)\cdot(x_j - x_1)\cdot\ldots\cdot(x_j - x_{j-1})\cdot(x_j - x_{j+1})\cdot\ldots\cdot(x_j - x_{n-1})\cdot(x_j - x_n)} \tag{9.15}$$

with

$$\binom{n}{m} = \begin{cases} \frac{n!}{m!\cdot(n-m)!} & \text{for } 0 \le m \le n \\ 0 & \text{for } 0 \le n < m \end{cases} \tag{9.16}$$

is the binomial coefficient

and

$$p_1 \in [0; n] \text{ without } \{j\} \wedge p_a \ne p_b. \tag{9.17}$$

References

1. Bronstein, Semendjajew; Taschenbuch der Mathematik; Verlag Nauka Moskau, 25. Auflage 1991
2. Wikipedia, Interpolationspolynom, 1. 3. 2021

Coordinate Transformations

10

Coordinate transformations are necessary whenever there is a need to switch from one coordinate system to another, for example, because representing a relationship is simpler in a different coordinate system. However, these transformations are not always straightforward and, above all, are not always universally valid.

10.1 Coordinate Systems

In principle, coordinate systems are used to describe points or sequences of points within a plane, a space, or a hyperspace. Coordinate systems are designed so that, depending on the number of dimensions of the plane, space, or hyperspace, at least a corresponding number of values describe the point or points; this set of values is referred to as the coordinates of the point or points.

Sequences of points in a coordinate system can be described using relations/functions. Depending on the number of independent and dependent variables, these relations/functions can be used to describe one- or multidimensional systems.

An example:

Consider a three-dimensional space. The coordinate system describing the points in this space is assumed to be an orthonormal system (see explanation in Sect. 10.1.1).

A function

$$z = 2 \cdot x + 3 \cdot y - 1 \text{ with } x, y, z \in R \tag{10.1}$$

describes a linear graph,

J. von Stackelberg, *The Mass of a Photon*, https://doi.org/10.1007/978-3-658-49408-7_10

while

$$
\begin{aligned}
&z = 2 \cdot x \\
&y = 4 \cdot x^2 - 3 \\
&\text{with } x, y, z \in R
\end{aligned}
\tag{10.2}
$$

describes a surface-like structure.

The dimensions in the plane, space, or hyperspace can be

- following a curved or straight line,
- following a closed line, e.g., a circle, an ellipse, or a polygon,
- following a sequence of layered surfaces,
- divided according to an irregular, a regularly irregular (e.g., logarithmic), or a uniformly stepped scale, etc.

embedded in the plane, space, or hyperspace.

The following sections describe some specific coordinate systems that are generally the most commonly used.

10.1.1 Orthonormal Cartesian Coordinate System

The orthonormal Cartesian coordinate system is a Cartesian coordinate system with mutually perpendicular coordinate axes, each divided into uniform and equal length intervals, and whose positive axes are arranged sequentially in the mathematically positive direction (counterclockwise in each case). For two-dimensional representation, two coordinate axes lie in the plane; for three-dimensional representation, three coordinate axes are positioned in space. Four-dimensional hyperspaces cannot be easily represented in three dimensions; in such cases, it depends on whether the space is an expanding system or a moving one (linear, rotational, etc.).

10.1.2 Polar Coordinate System

The polar coordinate system consists of concentric closed structures which, depending on the number of dimensions, form an object of $n - 1$ dimensions, e.g., a line for a circle, a surface for a sphere, etc. The remaining dimension represents the radius of the closed structures.

10.1.3 Coordinate Transformation within Cartesian Coordinates

10.1.3.1 Translation

In a translation within the Cartesian coordinate system, all points are shifted by the translation vector. The size and shape of structures remain unchanged.

10.1.3.2 Scaling

In scaling, starting from the origin of the coordinate system, the position vectors of all points in a structure are lengthened (or shortened) by the scaling factor. The shape of the structure remains unchanged; only its size changes.

A special case of scaling occurs when one dimension is transformed from linear to logarithmic scaling. Logarithmic scales are used when very large numerical ranges need to be represented with uniform relative accuracy. In terms of representation, this means that a function of the type $f(x) = e^x$ appears as a straight line in the graph.

10.1.3.2.1 Double Logarithmic Representation

It becomes particularly interesting when characteristic curves are represented using a double logarithmic coordinate system, for example, the tripping characteristic of a fuse (Fig. 10.1).

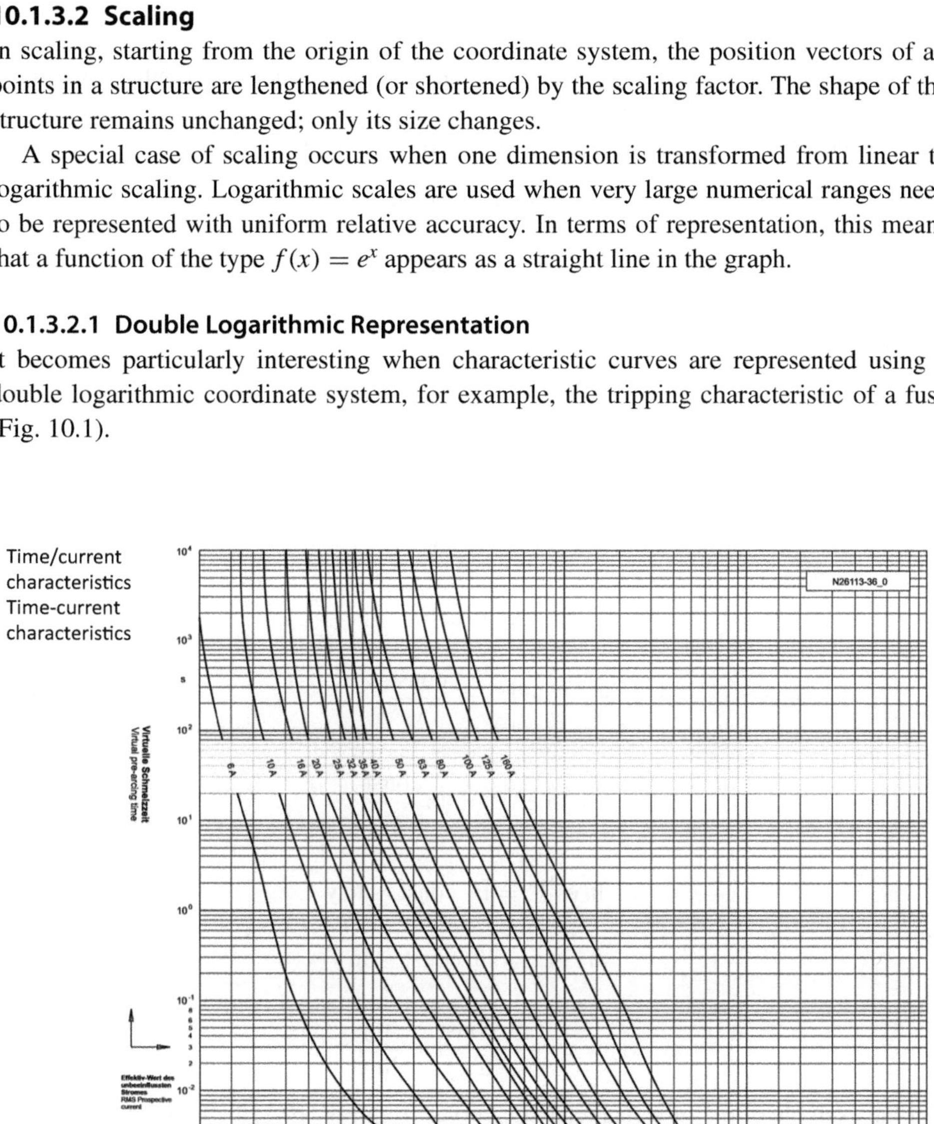

Fig. 10.1 Double logarithmic representation of time-current characteristic curves of NH fuses [1]

Tab. 10.1 Various value combinations for Eq. 10.3

a	1	1	2	1	100	1000
b	1	1	1	2	1	−1.5
c	0	10	0	0	0	0

If we consider the function

$$f(x) = a \cdot x^b + c \tag{10.3}$$

and assign values according to Table 10.1, the resulting graphs in a double logarithmic coordinate system are shown in Fig. 10.2.

If the equation $f(x) = a \cdot x^b + c$ is rearranged, we obtain

$$y \cdot x^a = \text{const} \tag{10.4}$$

$$x^a = \frac{\text{const}}{y} \tag{10.5}$$

Recall [2]:

$$a^x = e^{x \cdot \ln a} \tag{10.6}$$

$$x^a = \frac{\text{const}}{y} \tag{10.7}$$

$$x^a = e^{a \cdot \ln x} \tag{10.8}$$

$$e^{a \cdot \ln x} = \frac{\text{const}}{y} \tag{10.9}$$

$$\ln e^{a \cdot \ln x} = \ln \frac{\text{const}}{y} \tag{10.10}$$

$$\ln e^{a \ln x} = \ln \ \text{const} - \ln y \tag{10.11}$$

$$a \cdot \ln x = \ln \ \text{const} - \ln y \tag{10.12}$$

$$a = \frac{\ln \text{const}}{\ln x} - \frac{\ln y}{\ln x} \tag{10.13}$$

For the example of the fuse characteristic curves from Fig. 10.1, the following applies

$$S_1(100/30000); \quad S_2(1000/1,2) \tag{10.14}$$

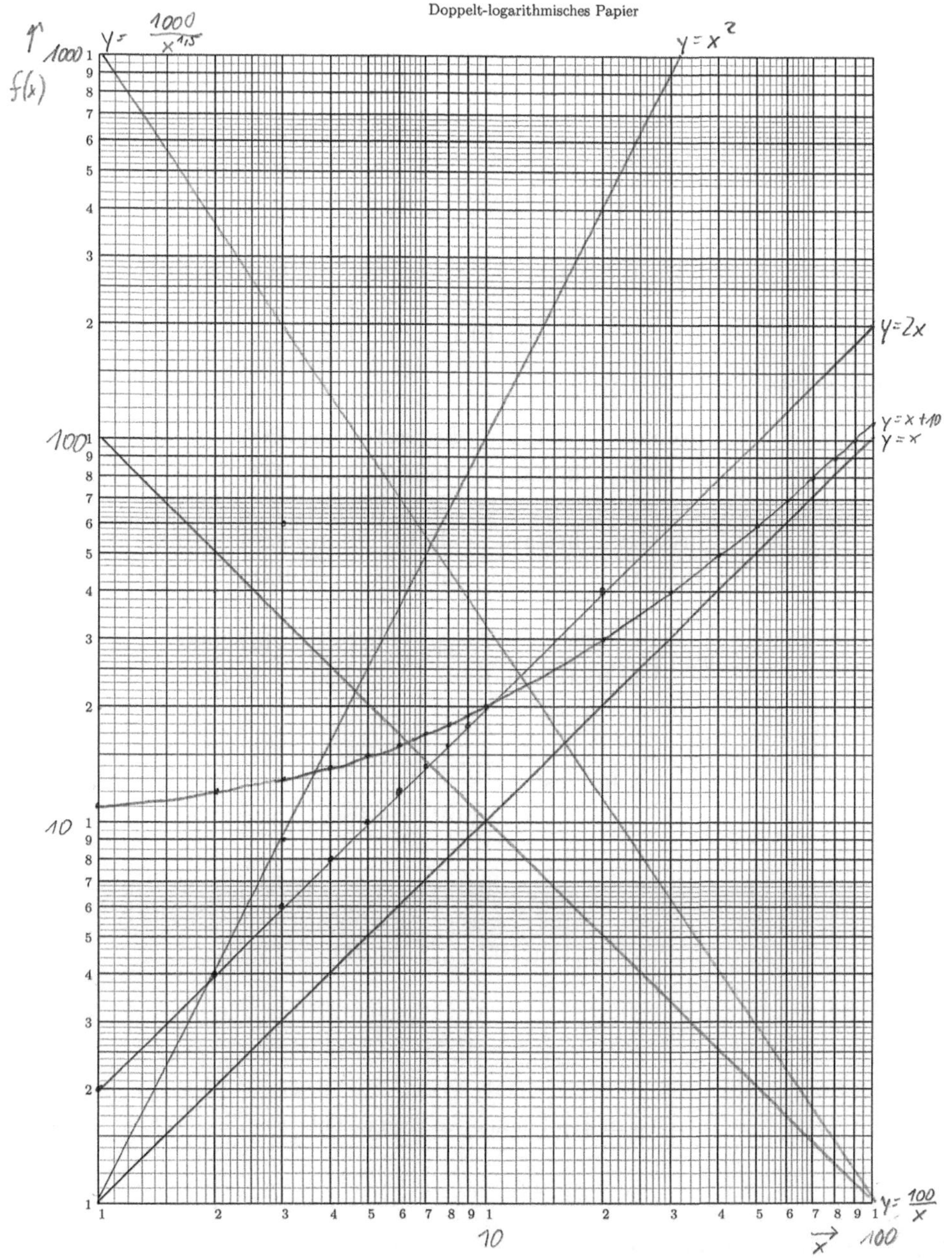

Fig. 10.2 Graphs for various functions of the type Eq. 10.3 in double logarithmic representation

Substituting into equation

$$a = \frac{\ln const}{\ln x} - \frac{\ln y}{\ln x} \tag{10.15}$$

to determine a and const:

$$\text{I} \quad a = \frac{\ln const}{\ln 100} - \frac{\ln 30000}{\ln 100} \tag{10.16}$$

$$\text{II} \quad a = \frac{\ln const}{\ln 1000} - \frac{\ln 1,2}{\ln 1000} \tag{10.17}$$

Equating I and II:

$$\frac{\ln const}{\ln 1000} - \frac{\ln 1,2}{\ln 1000} = \frac{\ln const}{\ln 100} - \frac{\ln 30000}{\ln 100} \tag{10.18}$$

$$\frac{\ln const}{\ln 1000} - \frac{\ln const}{\ln 100} = \frac{\ln 1,2}{\ln 1000} - \frac{\ln 30000}{\ln 100} \tag{10.19}$$

$$\ln const \cdot \left(\frac{1}{\ln 1000} - \frac{1}{\ln 100}\right) = \frac{\ln 1,2}{\ln 1000} - \frac{\ln 30000}{\ln 100} \tag{10.20}$$

$$\ln const = \frac{\frac{\ln 1,2}{\ln 1000} - \frac{\ln 30000}{\ln 100}}{\frac{1}{\ln 1000} - \frac{1}{\ln 100}} \tag{10.21}$$

$$a = \frac{\frac{\frac{\ln 1,2}{\ln 1000} - \frac{\ln 30000}{\ln 100}}{\frac{1}{\ln 1000} - \frac{1}{\ln 100}}}{\ln 100} - \frac{\ln 30000}{\ln 100} \tag{10.22}$$

$$const = e^{\frac{\frac{\ln 1,2}{\ln 1000} - \frac{\ln 30000}{\ln 100}}{\frac{1}{\ln 1000} - \frac{1}{\ln 100}}} \tag{10.23}$$

$$a = 4.4; \quad const = 1.875 \cdot 10^{13} \tag{10.24}$$

$$y \cdot x^{4.4} = 1.875 \cdot 10^{13} \tag{10.25}$$

This means that in the time range from 10 ms to 100 s of disconnection time, the current in the current-time characteristic curve affects the disconnection behavior to the 4.4th power relative to time. In general, the energy depends on the square of the current. The 4.4th power indicates that the fuse receives significantly less heat from the current before the fuse wire opens than would be required by the melting heat. This phenomenon occurs because the fuse wire is pre-tensioned, for example by a tension spring, and breaks before reaching the melting point.

10.1.3.3 Rotation

In a rotation, there is a point about which the coordinate system rotates, and an angle that describes the rotation.

10.1.4 Coordinate Transformation from Cartesian to Polar Coordinate System

For simplicity, let us consider a two-dimensional Cartesian coordinate system and a two-dimensional polar coordinate system whose origins coincide.

The representation of a point in the Cartesian coordinate system as a position vector has the following structure:

$$\vec{P}_{kart} = \begin{pmatrix} x_P \\ y_p \end{pmatrix} \tag{10.26}$$

This point has the following form in the corresponding polar coordinate system:

$$\vec{P}_{pol} = \begin{pmatrix} r \\ \varphi \end{pmatrix} \tag{10.27}$$

The transformation rules are as follows

$$r = \sqrt{x_P^2 + y_P^2} \tag{10.28}$$

$$\varphi = \arctan \frac{y_P}{x_P} \tag{10.29}$$

or

$$x_P = r \cdot \cos \varphi \tag{10.30}$$

$$y_P = r \cdot \sin \varphi \tag{10.31}$$

In the Cartesian coordinate system, there are also vectors, i.e., directed arrows from one point to another:

$$\vec{v_k} = \begin{pmatrix} x_V \\ y_V \end{pmatrix} \tag{10.32}$$

If two points $A(x_A/y_A)$ and $B(x_B/y_B)$ are given, the connecting vector $\overrightarrow{AB}$ is calculated as follows:

$$\overrightarrow{AB} = \overrightarrow{v_B} - \overrightarrow{v_A} \tag{10.33}$$

$$\overrightarrow{AB} = \begin{pmatrix} x_B \\ y_B \end{pmatrix} - \begin{pmatrix} x_A \\ y_A \end{pmatrix} \tag{10.34}$$

$$\overrightarrow{AB} = \begin{pmatrix} x_B - x_A \\ y_B - y_A \end{pmatrix} \tag{10.35}$$

The graph of this vector is a directed arrow from point A to point B. The graph is a straight line running from point A to point B.

The coordinate transformation of the two points A and B from Cartesian to polar coordinates yields:

$$A_P = \begin{pmatrix} \sqrt{x_A + y_A} \\ \arctan \frac{y_A}{x_A} \end{pmatrix} \tag{10.36}$$

$$B_P = \begin{pmatrix} \sqrt{x_B + y_B} \\ \arctan \frac{y_B}{x_B} \end{pmatrix} \tag{10.37}$$

The vector AB runs along the diagonal of a rectangle whose sides are parallel to the coordinate axes.

Accordingly, the resulting vector would be

$$\overrightarrow{v_{Ep}} = \overrightarrow{v_{Bp}} - \overrightarrow{v_{Ap}} \tag{10.38}$$

$$\overrightarrow{v_{Ep}} = \begin{pmatrix} \sqrt{x_B + y_B} \\ \arctan \frac{y_B}{x_B} \end{pmatrix} - \begin{pmatrix} \sqrt{x_A + y_A} \\ \arctan \frac{y_A}{x_A} \end{pmatrix} \tag{10.39}$$

$$\overrightarrow{v_{Ep}} = \begin{pmatrix} \sqrt{x_B + y_B} - \sqrt{x_A + y_A} \\ \arctan \frac{y_B}{x_B} - \arctan \frac{y_A}{x_A} \end{pmatrix} \tag{10.40}$$

Without addressing every special case, this leads to

$$\overrightarrow{v_{Ep}} = \begin{pmatrix} \sqrt{\left(\sqrt{x_B + y_B} - \sqrt{x_A + y_A}\right)^2} \\ \arctan \left(\frac{\frac{y_B}{x_B} - \frac{y_A}{x_A}}{1 + \frac{y_B}{x_B} \cdot \frac{y_A}{x_A}} \right) \end{pmatrix} \tag{10.41}$$

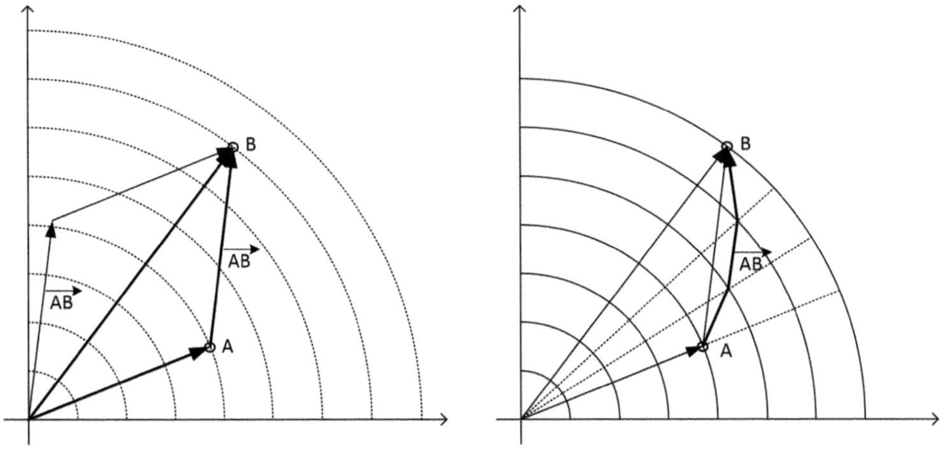

Fig. 10.3 Construction of a vector in the polar coordinate system from point A to point B; in the left coordinate system (Cartesian), the vector AB can be easily constructed and, as a free vector, can be shifted anywhere without changing its values; in the right coordinate system (polar), the vector AB has a spherical contour; the construction is carried out by dividing the space between the two points into segments of equal angle and radius, and the vector is drawn diagonally from intersection point to intersection point; since the distance between two radial lines is smaller at a small radius than at a large radius, a curved line is formed

$$\overrightarrow{v_{Ep}} = \begin{pmatrix} \sqrt{x_B + y_B + x_A + y_A - 2 \cdot \sqrt{x_B + y_B} \cdot \sqrt{x_A + y_A}} \\ \arctan\left(\frac{\frac{y_B}{x_B} - \frac{y_A}{x_A}}{1 + \frac{y_B}{x_B} \cdot \frac{y_A}{x_A}}\right) \end{pmatrix} \quad (10.42)$$

Fundamentally, the question arises as to how a connecting line between two points should be represented in a polar coordinate system: as a "straight" line in the sense of the Cartesian coordinate system, or as a "spherical diagonal" within the concentric circular segments spanned by the two angles defined by points A and B?

An auxiliary construction provides the answer (see Fig. 10.3).

While vectors can be elegantly manipulated in the Cartesian coordinate system, polar coordinate systems, although they offer more interesting geometric forms, are essentially limited in their practical use to position vectors.

References

1. NH-Katalog 2019; Siba GmbH
2. Bronstein, Semendjajew; Taschenbuch der Mathematik; Verlag Nauka Moskau, 25. Auflage 1991

GPSR Compliance

The European Union's (EU) General Product Safety Regulation (GPSR) is a set of rules that requires consumer products to be safe and our obligations to ensure this.

If you have any concerns about our products, you can contact us on ProductSafety@springernature.com

In case Publisher is established outside the EU, the EU authorized representative is:

Springer Nature Customer Service Center GmbH
Europaplatz 3
69115 Heidelberg, Germany

Batch number: 10683872

Printed by Printforce, the Netherlands